天地
中和

Harmony of Heaven and Earth
Cultural Heritage of Beijing Central Axis

北京中轴线文化遗产

吕舟 主编

Chief Editor LYU Zhou

北京出版集团
北京出版社

《天地中和——北京中轴线文化遗产》编委会
Editorial Board of *Harmony of Heaven and Earth: Cultural Heritage of Beijing Central Axis*

主编: 吕舟
Chief Editor: LYU Zhou

执行主编: 初迎霞
Executive Editor: CHU Yingxia

副主编: 波音、孙燕
Associate Editors: BO Yin, SUN Yan

编委: 郑楚晗、邵龙飞、张娜、田芯祎、邓阳雪、王勇、李琴
Editors: ZHENG Chuhan, SHAO Longfei, ZHANG Na, TIAN Xinyi, DENG Yangxue, WANG Yong, LI Qin

技术指导单位: 清华大学建筑设计研究院有限公司
Technical Advisor: Architectural Design & Research Institute Tsinghua University Co., Ltd

英文翻译: 解立
Translator: XIE Li

「致中和，天地位焉，万物育焉。」

When the Mean and Harmony are actualized, Heaven and Earth are in their proper positions, and the myriad things are nourished.

2013年航拍钟鼓楼地段（马文晓 摄）
Aerial view of Bell and Drum Towers Area in 2013 (Photo by Ma Wenxiao)

航拍故宫（龙脉航拍／供图）
Aerial view of the Forbidden City (Provided by Longmai Aerial Image)

航拍天安门广场 (马文晓 / 摄)
Aerial view of Tian'anmen Square (Photo by Ma Wenxiao)

航拍永定门（龙脉航拍／供图）
Aerial view of Yongdingmen Gate (Provided by Longmai Aerial Image)

故宫汉白玉石阶 (李少白／摄)
White marble steps in the Forbidden City (Photo by Li Shaobai)

编者说明

Editor's Note

北京中轴线是一条全长7.8公里的轴线,北起钟鼓楼,向南经万宁桥、景山、故宫至永定门,纵贯北京老城南北,是由中国古代皇家建筑、城市管理设施和居中历史道路、现代公共建筑和公共空间共同构成的城市历史建筑群,也是世界上最长的一条都城中轴线。

2009年北京中轴线申报世界遗产工作开始推进。北京中轴线申遗,不仅仅是希望跻身于《世界遗产名录》,更重要的是通过申遗促进整个北京老城的保护,规范北京历史文化价值的系统表达,促进城市管理水平的提升,改善市民的生活环境,促进城市的可持续发展。

目前越来越多的人参与到北京中轴线申遗工作中来,我们与申遗文本团队合作,适时推出《天地中和——北京中轴线文化遗产》一书。书名"天地中和",语出《中庸》"致中和,天地位焉,万物育焉",可以理解为"中和定,乾坤立,天地位,万物育,众生各得其所"。北京中轴线正是千年传承的"天地中和"文化精神的集中体现。

本书的图书结构和价值阐释,在充分尊重申遗文本的基础上,广泛参考了其他重要文献材料。全书分为五个部分,第一部分是吕舟先生撰写的北京中轴线概述及其遗产价值;第二到第五部分对北京中轴线的主要遗产段落"钟鼓楼—地安门段落""地安门—天安门段落""天安门广场建筑群""正阳门—永定门段落"分别进行了展示。

全书图文并茂,汇集了李少白、马文晓、朱雨生等众多优秀摄影家的精彩作品,辅之以重要的历史舆图和老照片;图书整体装帧设计提取了北京中轴线布局及建筑的相关文化元素,以期内容与形式互为映照,更好地呈现北京中轴线的历史、文化内涵与风采。此外,本书对内文标题、图注及重要章节段落做了英文翻译,方便国际友人更便捷地通过图册一览北京中轴线的概貌。

北京中轴线历史文化内容深厚,形态丰富多彩,编者虽勉力为之,亦难免瑕疵和不足,恳请方家不吝指正。编者对为本书提供多方面支持的专家、学者及同仁致以诚挚谢意!感谢大家借由此图册的编辑共同为北京中轴线申遗所贡献的一份力量!

Beijing Central Axis is a 7.8-kilometer-long axial line that runs through the old city of Beijing, starting from the Bell and Drum Towers in the north and extending southward through the Wanning Bridge, the Jingshan Hill and the Imperial Palace down to the Yongdingmen Gate at the south end. It is an ensemble of city historical architectures that comprises imperial buildings from ancient China, city management facilities, the central historical roads, and modern structures and space for public use. Beijing Central Axis, as a capital-city axis road, tops the world in its length.

Beijing Central Axis began its journey of World Heritage nomination in 2009. The campaign not only aims at the inscription of Beijing Central Axis on the *World Heritage List*, but more importantly, the integral protection of the old city of Beijing, improved systematic expression of Beijing's historical and cultural values, better management of the city, enhanced living environment of citizens, and sustainable development of the city.

As more and more people have participated in work related to the World Heritage nomination of Beijing Central Axis, we hope to take this opportunity and work closely with the application text compilation team towards the timely publication of the book *Harmony of Heaven and Earth: Cultural Heritage of Beijing Central Axis*. The phrase "Harmony of Heaven and Earth" in the name of the book originates from a quote from the ancient Chinese classic *Zhongyong* (or *The Doctrine of the Mean*) that "When the Mean and Harmony are actualized, Heaven and Earth are in their proper positions, and the myriad things are nourished". The quote can be interpreted as that when the states of "Centrality" and "Harmony" have been achieved, Heaven and Earth will be established and in the right position, and all things will be nourished and properly placed. Beijing Central Axis is the embodiment of the quintessential cultural spirit of "harmony between heaven and earth" that has been passed down in thousands of years.

In terms of the structure and value interpretation, this book fully respects the application text, and in addition, extensively refers to other important literature. The book contains five parts, with the opening part written by Mr. LYU Zhou on the overview of Beijing Central Axis and its heritage value. The other four parts demonstrate respectively such major heritage sections of Beijing Central Axis as the "Bell and Drum Towers-Di'anmen Section", the "Di'anmen Gate-Tian'anmen Gate Section", the "Tian'anmen Square Complex", and the "Zhengyangmen Gate-Yongdingmen Gate Section".

The book is brilliantly presented in both literary composition and accompanying pictures. It brings together the wonderful works of many outstanding photographers including Li Shaobai, Ma Wenxiao and Zhu Yusheng, complemented by important historical maps and old photographs. The bookbinding and layout design employ a selection of cultural elements characterizing the plan of Beijing Central Axis and its composing architectures, in order that the content and form echo into better presentation of the history, cultural connotation and graceful bearing of Beijing Central Axis. In addition, English translation is given to the section headings, captions and important chapters and paragraphs, so that international friends can access an overview of Beijing Central Axis through the illustrated book.

As Beijing Central Axis contains historical and cultural contents of such volume, in such depth and of such a variety of forms, that despite our best efforts, there must still be inevitable flaws and defects. In that case, we invite all those with expertise to correct us anytime without hesitation. We would like to extend our gratitude to the experts, scholars and colleagues for their all-round support to this book and their contribution, via this illustrated book, to the World Heritage application of Beijing Central Axis.

目录 CONTENTS

故宫角楼一角（朱雨生／摄）
Part of the Corner Tower of the Forbidden City (Photo by Zhu Yusheng)

Echoes of a Millennium

北京中轴线
Beijing Central Axis

天地中和

Harmony of Heaven and Earth

从历史传统中显现的北京中轴线遗产及其价值

Heritage and Value of Beijing Central Axis from the Perspective of Historical Traditions

吕舟（LYU Zhou）

什么是北京中轴线

北京是中国现存最为完整的古代都城。北京中轴线作为北京老城中最重要、最高等级建筑构成的建筑群和空间形态，随着申报世界遗产和相关保护、整治以及文物腾退等工作的开展，已经成为一个受到社会广泛关注的遗产对象。

20世纪40年代，侯仁之先生曾经在他的博士论文《北平历史地理》中提到北京中轴线：

"在确定城墙时，大致遵循如下原则：第一，在积水潭东北岸建钟楼，以确定大城的几何中心，并于稍南建鼓楼。第二，钟鼓楼坐落在皇城中轴线延长线上，因此这条中轴线与东西城墙等距，钟楼还与南北城墙等距。"

该论文还提到："到1420年北京城的重建按计划完成时，人们看到，新规划的所有重要建筑都以这条南起大城前门，北至钟鼓二楼的直线为中线东西均衡对称分布。这种沿着中轴线布局的几何图案之美尤其引人注目，这一布局一直保留至今。"

侯仁之先生把北京中轴线视为元大都（汗八里）及之后明代北京城市规划和布局的一种方法，这种方法构成了北京城引人注目的"图案之美"。

梁思成先生在1951年发表的《北京——都市计划的无比杰作》一文中对北京中轴线做了极富感情色彩的描述：

"大略地说，凸字形的北京，北半是内城，南半是外城，故宫为内城核心，也是全城布局重心，全城就是围绕这中心而部署的。但贯通这全部部署的是一根直线。一根长达八公里，全世界最长，也最伟大的南北中轴线穿过了全城。北京独有的壮美秩序就由这条中轴线的建立而产生。前后起伏左右对称的体形或空间的分配都是以这中轴为依据的。气魄之雄伟就在这个

注：中英文不完全对应。Please note that the English translation is not the whole text.

『大略地说，凸字形的北京，北半是内城，南半是外城，故宫为内城核心，也是全城布局重心，全城就是围绕这中心而部署的。但贯通这全部部署的是一根直线。一根长达八公里，全世界最长，也最伟大的南北中轴线穿过了全城。北京独有的壮美秩序就由这条中轴线的建立而产生。』

——梁思成，《北京——都市计划的无比杰作》，1951年

In his article *Beijing: An Exceptional Masterpiece of Urban Planning* published in 1951, Liang Sicheng gave an emotional description of Beijing Central Axis that reads:

" There is a straight line that connects all these layout arrangements. This is an eight-kilometer-long central axis that runs through the entire city from south to north, hailed as the longest and greatest of its kind in the world. This central axis gives rise to the majestic spatial order that is unique to Beijing and governs arrangement of all architectural complexes and spatial forms symmetrically distributed along it. The grandeur of the city rests on this central axis that extends from south to north."

What is Beijing Central Axis?

Beijing maintains the most complete ancient capital city that survives till today in China. Beijing Central Axis is the architectural ensemble and spatial form comprising the most important buildings of the highest grade in the old city of Beijing. With its ongoing UNESCO World Heritage nomination with related heritage conservation and environmental rehabilitation, Beijing Central Axis attracts ever-increasing attention from the society.

There was a mention of Beijing Central Axis in Hou Renzhi's doctoral thesis *An Historical Geography of Peiping* written in the 1940s that reads:

"The principles which were applied to decide the site of the outer rampart may conceivably have been the following: First, a bell tower was built on the northeast bank of the lake Chi Shui T'an (积水潭) to fix the geometrical centre of the great city. A little south of the bell tower, a drum tower was also built. Second, both towers were in line with the central axis of the emperor's palace. Consequently this line, the central axis of the emperor's palace, was equidistant from the east and west walls of the great city, and the bell tower was equidistant from the north and south walls."

"Thus by the year 1420, when the plan of the redevelopment of the city was accomplished, we find that all the important features of the new plan were centred on the straight line, in well-balanced symmetrical disposition, between the front gate of the great city in the south and the bell tower and drum tower in the north. The beauty of the geometrical pattern of the layout along this line is most striking, and this has been preserved up to the present day." He also mentioned.

Hou Renzhi perceived Beijing Central Axis as a method of city planning and layout for Dadu of the Yuan Dynasty (Khanbaliq the Great Capital) and the ensuing capital of the Ming Dynasty. This method presents a fascinating "beauty of patterns" for the city of Beijing.

In his article *Beijing: An Exceptional Masterpiece of Urban Planning* published in 1951, Liang Sicheng gave an emotional description of Beijing Central Axis that reads:

"Roughly, the shape of Beijing resembles the Chinese character "凸" (tu), with the Inner City in the north and the Outer City in the south. The whole city is designed and arranged around the Forbidden City located at the core of the Inner City. There is a straight line that connects all these layout arrangements. This is an eight-kilometer-long central axis that runs through the entire city from south to north, hailed as the longest and greatest of its kind in the world. This central axis gives rise to the majestic spatial order that is unique to Beijing and governs arrangements of all architectural complexes and spatial forms symmetrically distributed along it. The grandeur of the city rests on this central axis that extends from south to north."

In fact, it is Liang Sicheng's description that inspires us to nominate Beijing Central Axis to the *World Heritage List*. Just as Liang Sicheng said, "This is a great heritage and the most valuable property for our people. Are there anyone who do not think so?"

Over the past 70 years since Hou Renzhi and Liang Sicheng published their elaborations on Beijing Central Axis, many changes have taken place in Beijing. Some buildings on the Central Axis have disappeared, while some new additions have emerged in the process of urban construction and development and some disappeared buildings have been excavated or reconstructed. So, how to describe Beijing Central Axis from the perspective of cultural heritage in the cotemporary times?

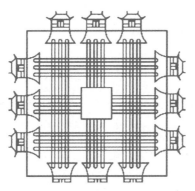

宋代聂崇义根据《周礼·考工记》绘制的理想都城范式模型
Design of an ideal capital city drawn by Nie Chongyi of the Song Dynasty based on the Ancient book of
Kao Gong Ji (Artificer's Record) in *Zhou Li (Rites of Zhou)*

南北引伸，一贯到底的规模。我们可以从外城最南的永定门说起，从这南端正门北行，在中轴线左右是天坛和先农坛两个约略对称的建筑群；经过长长一条市楼对列的大街，到达珠市口的十字街口之后才面向着内城第一个重点——雄伟的正阳门楼。在门前百余公尺的地方，拦路一座大牌楼，一座大石桥，为这第一个重点做了前卫。但这还只是一个序幕。过了此点，从正阳门楼到中华门，由中华门到天安门，一起一伏、一伏而又起，这中间千步廊(民国初年已拆除)御路的长度，和天安门面前的宽度，是最大胆的空间的处理，衬托着建筑重点的安排。这个当时曾经为封建帝王据为己有的禁地，今天是多么恰当地回到人民手里，成为人民自己的广场！由天安门起，是一系列轻重不一的宫门和广庭，金色照耀的琉璃瓦顶，一层又一层地起伏峋峙，一直引导到太和殿顶，便到达中线前半的极点，然后向北，重点逐渐退削，以神武门为尾声。再往北，又'奇峰突起'地立着景山做了宫城背后的衬托。景山中峰上的亭子正在南北的中心点上。由此向北是一波又一波的远距离重点的呼应。由地安门，到鼓楼、钟楼，高大的建筑物都继续在中轴线上。但到了钟楼，中轴线便有计划地，也恰到好处地结束了。中线不再向北到达墙根，而是将重点平稳地分配给左右分立的两个北面城楼——安定门和德胜门。有这样气魄的建筑总布局，以这样的规模来处理空间，世界上就没有第二个！"

事实上，也恰恰是由于梁思成先生这段关于北京中轴线的表述，引发和激励了我们今天要把这样一处遗产列入《世界遗产名录》，将它作为一处能够令整个人类社会为之自豪的遗产呈现在世界面前。就如梁思成先生所说的："这是一份伟大的遗产，它是我们人民最宝贵的财产，难道还有人感觉不到吗？"

侯仁之、梁思成两位先生发表的北京中轴线的论述已经过去七十年，在这七十年间，北京城发生了许多变化。北京中轴线的一些建筑已湮没在历史长河之中，不复存在；也有一些新的部分在城市建设和发展过程中呈现出来；还有一些消失的建筑又被重新发掘出来，被重建。那么，从当代文化遗产的视角，北京中轴线又是什么样的呢？

在今天文化遗产保护的语境中，北京中轴线是北起钟楼、鼓楼，南至永定门，由位于城市中心线上所有重要的建筑及两侧对称布局的太庙、社稷坛和天坛、先农坛四组坛庙建筑群构成的庞大建筑群。同样，被这些建筑、建筑群围合而成的空间、联系这些建筑的街道也是北京中轴

《周礼·考工记》中营国制度对城市内重要建筑的分布做了规定：朝堂应当位于城市的前部（南部），市场应当位于城市的后部（北部），宗庙应当位于城市的左半部（东部），而社稷坛则应位于城市的右半部（西部），即『左祖右社，面朝后市』。

Kao Gong Ji (Artificer's Record) in Zhou Li (Rites of Zhou) regulated the general layout of a capital city that features "the court in the front and the marketplace in the back, the ancestral temple on the left and the altar of land and grain on the right".

Beijing Central Axis is an immense architectural ensemble comprising all the important buildings on or along both sides of it, which stretches from the Bell and Drum Towers in the north to the Yongdingmen Gate Tower in the south. In addition, spaces encircled by those buildings and the streets connecting them are also comprising parts of Beijing Central Axis. It encompasses the central area of Beijing's old city and forms its skylines that are undulating, varying and full of rhythms. Beijing Central Axis brings together the most important buildings dating from the Yuan Dynasty until present-day China and conveys the traditional Chinese worldview highlighting "respect for the central". It displays a strict hierarchical relationship among key buildings along the axis line and on its sides, and extends such relationship to street layouts of the entire old city or even the national territory as a whole.

Beijing Central Axis presents a physical expression of the relationship between man and nature as conceived by philosophical doctrines in ancient China, embodies the traditional pursuit of order, and epitomizes Chinese people's practices to build an ideal order and an ideal homeland in a disordered real world. It is a crystallization of the core spirit of Chinese cultural traditions.

Construction of capital cities in ancient China and creation of Beijing Central Axis

The Yuan Dynasty saw dramatic changes in Chinese capital city planning system. According to surviving historical literature, after choosing Jishuitan Lake as the core area of Dadu of the Yuan Dynasty, Liu Bingzhong, the designated architect, constructed the "central platform" as the focal site of urban construction, determined boundaries of the capital on four sides, and finalized the southward axis starting from the "central platform". Imperial palaces were then arranged in the south section of the axis. The capital layout planning had not been recorded in any literature from previous dynasties. Liu Bingzhong did not place the Palace City at the capital's geometric center. It was a layout plan with unprecedent innovation. Perhaps because Kublai Khan had become the ruler of a massive empire, when planning the Great Capital, Liu Bingzhong not only followed some planning practices for the earlier Upper Capital, but also drew upon from the form of the imperial capital described in the ancient book of *Kao Gong Ji (Artificer's Record)* in *Zhou Li (Rites of Zhou)*, such as the city layout that features "the court in the front and the marketplace in the back, the ancestral temple on the left and the altar of land and grain on the right". This is also China's only surviving example that exhibits the city form as described in *Kao Gong Ji*. In Dadu, the Great Capital, the city's central axis starting from the "central platform" and running across the Lizheng Gate (the south gate of the city) connected the southern part of the capital, the Palace City, the Imperial City and the marketplace. It gave rise to a new form of imperial capital that originates from the capital city construction system of the Western Zhou Dynasty and complies with the Confucian philosophy of order.

When Nanjing was set as the imperial capital in the early Ming Dynasty, buildings in the Palace City of Dadu of the Yuan Dynasty were demolished and the open stretch of space in its north abandoned, which significantly reduced the size of the city. Following Emperor Yongle's decision to relocate the capital from Nanjing to Beijing, construction of the Inner City, the Palace City and the Imperial City began in 1407 in Beijing. The complete layout of royal palaces

线的组成部分。北京中轴线是北京老城的中心区域,它构成了北京老城起伏变化、富有韵律感的城市天际线。北京中轴线汇集了从元代到今天中国最重要的建筑,表达中国传统的"以中为尊"的世界观。北京中轴线展现了位于这一轴线上和两侧的重要建筑形成的严格等级秩序关系,并将这种秩序关系扩大到整个北京老城的街巷格局乃至更广大的国土范围。

从其所反映的观念看,北京中轴线是中国传统哲学中对于天道与人间世界关系认知的物质表达,反映了中国传统观念中对秩序的追求,呈现了中国人在无序的外部环境中构建理想秩序、建立理想家园的实践,是中国文化传统的核心精神的体现。

中国古代都城建设与北京中轴线的形成

在公元前1046年周朝建立之初,由于周王朝大封诸侯,中华文明出现了第一个城市建设的高峰,同时形成了城市的等级制度和布局范式。这种等级制度和布局范式通过春秋末期的文献被记载和传承下来,其中《周礼·考工记》所记载的营国制度反映了在西周分封制度影响下,以等级制度为核心的城市建设规划体系。在这个体系中,城市的规模是根据所居住的统治者的身份确定的:天子的国都方九里,诸侯方七里,大夫方五里。从西周分封制度的角度,这一体系是秩序的组成部分,具有明显的封建等级制度在城市空间形态、规模上延续的特征,是社会秩序在空间上的表达。营国制度还对城市内重要建筑的分布做了规定:朝堂应当位于城市的前部(南部),市场应当位于城市的后部(北部),宗庙应当位于城市的左半部(东部),而社稷坛则应位于城市的右半部(西部),即"左祖右社,面朝后市"。

如果说《周礼·考工记》反映了自西周以来分封制度下对城市规模和形态的规定,《管子》则体现出在春秋战国时期原有礼仪秩序解体过程中更加基于城市基本功能的规划思想。其核心是根据城市的人口规模确定城市的尺度,根据城市所在的环境确定城市的形态和位置,根据城市的功能确定城市中各部分的位置。这两种似乎相悖的城市规划理念反映了中国城市发展过程中的两个方面,而这两个方面在秦、汉之后的中国城市建设中又同时影响着建设活动。

秦始皇在建立统一大帝国的同时,对都城咸阳进行了大规模的建设。据文献描述,咸阳的宫殿模仿星象,将咸阳宫视为天帝居住的紫微垣,其他宫室按照星座的分布环绕咸阳宫布局,穿越咸阳的渭河则被视为横跨天穹的银河,形成了一种模仿天体的宏大城市形态。

汉代的长安城在城市形态上一定程度延续了秦代模仿天体的特征,城市呈不规则布局,但将东市、西市置于由未央宫、桂宫和北宫构成的宫殿建筑群之后,在一定程度上呈现了"面朝后市"的城市形态。汉长安城内集中了大量的宫殿建筑群,从功能上看,似乎更多反映了"筑城以卫君"的城市功能。汉代长安城融合了部分周代城市制度,又结合了秦代的城市形态,具有承前启后的重要意义。值得注意的是,西汉末年王莽新朝对长安的城市布局进行了调整,所建长安南郊的祭祀建筑群落对后期城市制度有重要影响。

但无论在秦咸阳,还是汉长安,从现有资料看都未出现影响整个城市布局的城市中轴线形态。

东汉洛阳不同于汉长安的城市形态,它规划设置了纵贯城市南北的两个大宫殿区,其中南部宫殿区有从前殿直通城门,并达城外南郊的灵台、明堂等祭祀礼仪区域的大道。

东汉末期的曹魏邺城一改汉代将宫室置于城市南端的布局方式,将宫室设置于城市北半部,从宫廷区主殿延伸出的大道穿过城市南部,与南北主要干道融为一体,在这一主干道两侧布置衙署等政府机构,形成对城市整体布局的控制。这就是城市中轴线的雏形。

(1)

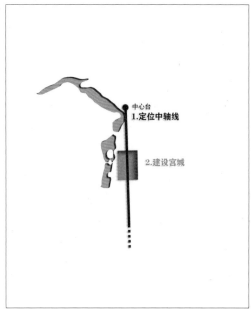

(2)

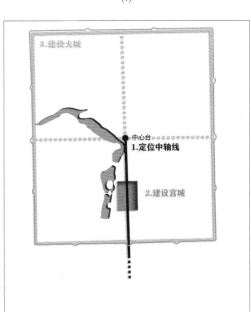

(3)

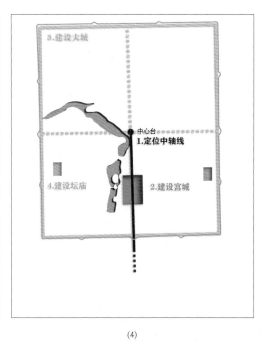

(4)

元大都规划与建设时序示意图(清华大学建筑设计研究院北京中轴线申遗文本项目组/供图)
Chronological diagram on the planning and construction of Dadu of the Yuan Dynasty (Provided by THAD Team for the Nomination Dossier of Beijing Central Axis)

that developed during the construction of the capital Nanjing was then fully applied to the contruction of Beijing. As the Ming Dynasty capital was relocated southward from the site of Dadu of the Yuan Dynasty, the original central axis in the city's southern part became the Beijing Central Axis that ran through the entire city. While following the layout of "the court in the front and the marketplace in the back" and setting the Bell and Drum Towers as city management facilities at the north end of the axis line, city planners of the Ming Dynasty arranged the Imperial Ancestral Temple and the Altar of Land and Grain on east and west sides of the central axis respectively, close to the southern section of the Imperial City, giving rise to a more compact and stricter layout of the central axis. On both sides of the southern extension of the central axis were the Altar of Heaven and the Earth and the Altar of Mountains and Rivers to offer sacrifices to various

北魏洛阳延续了曹魏邺城的布局方式,宫室同样设置在城市北半部,从太极殿向南的大道出宫城闾阖门,经城市南北主要大街铜驼街至内城宣阳门,出宣阳门进入外郭城,经永桥跨洛河,向南直达圜丘。铜驼街东西两侧设衙署。这一城市轴线尽管略微偏西,并不处于内城的几何中心,但却具有城市主干道的重要性,在整个城市形态中处于主导地位。

以纵贯坐北朝南的大朝正殿的南向御路作为城市中心干道和对称轴的都城规划,在隋唐长安城的建设中布局更加严正均衡。

唐代的长安城是在隋代大兴城的基础上延续而成的。唐长安城将宫城置于城市北端中心,宫城以南为皇城。宫城分为东侧的东宫、中部的太极宫和西侧的掖庭宫三个部分。太极宫南向正中的城门为承天门,从承天门穿过皇城正门朱雀门,再到长安城南端中心的明德门,这一城市中心的主干道形成几何意义上的城市中轴线。长安城以这条轴线为核心,东西对称布置街道、市场、里坊,形成了宏大、严谨的城市格局。这种都城形态也影响了渤海上京城及日本平城京和平安京的规划。

宫室设在城市北端,呼应了传统的帝王应居住在对应"天极"或"紫微垣"位置的观念。宽阔笔直的中央干道、两侧严谨对称的建筑布局形成了宏伟、庄严的城市面貌特征。这种规划布局决定了中国城市发展成熟期都城的形态特征。

北宋东京是在后周都城的基础上发展而成的,受到许多旧有条件的限制,但依然形成了从大内正门宣德门经过州桥,经过内城正南门朱雀门,再跨越龙津桥,直抵外城正南门南薰门的城市主干道(或城市中轴线)。这种由宫殿、城门和干道构成的城市轴线被赋予了都城气象和帝王品格的象征意义。从城市形态的角度,这种以大朝殿为起点,以穿越大朝殿以南城市区域的皇家大道为中轴线对称布局的中国都城格局,从曹魏邺城开始到隋唐长安发展至顶峰。而到宋代,随着经济的发展,都城建设突破里坊制度的束缚,无论是汴梁还是临安的城市形态都发生了变化。辽金时期的都城则更是以唐长安为代表的都城形态的余音。

元代是中国古代都城规划体系发生巨大变化的时期。在兴建大都城之前,1256年忽必烈兴建开平府,1263年将开平府升格为"上都",成为其都城。上都城的格局与汉唐以来中原地区的都城存在明显差别,这也许与兴建之初它的性质并非都城相关。整个城市分为三个部分:大城的东南部为皇城,皇城中部靠北为宫城;大城西南部为城市区域;大城北半部则为空阔的"御苑"区域。皇城和宫城存在明显的轴线对位遗迹,形成了中轴线的形态,但这一中轴线仅为皇城和宫城的中轴线,而并非整个上都城的中轴线。上都城与后来元大都的规划者都是忽必烈的宠臣刘秉忠。

在建设开平府十年后,忽必烈决定定都北京,1267年委派刘秉忠负责大都城的规划建设。据遗留的相关文献记载,刘秉忠在确定以积水潭水系为城市核心区域之后,首先堆筑"中心台"作为城市建设的中心点,之后确定了大都城的四至范围,确定了以中心台为起点的南向轴线,并在这一轴线的南部布置大内宫殿。这一先建中心台、确定城市轴线,再进行城市主要功能区划分的规划方法,在之前历代都城建设的相关文献中都未见记载。而且刘秉忠也未将宫城(大内)放在城市几何中心位置,这无疑是一种全新的规划方法。或许是由于忽必烈已经成为一个庞大帝国的帝王,刘秉忠在规划大都城时,除了延续上都规划的一些手法,还明显附会了《周礼·考工记》的国都形态,如"面朝后市,左祖右社"的城市布局。这也是中国现存古代都城中唯一完整展现《周礼·考工记》城市形态的案例。在大都城中,自中心台穿过大城南门丽正门的城市中轴线串起了大都城的南半部分——宫城、皇城和市场,形成新的城市形态,构建了一个源自于西周王都制度、

至正年间的元大都示意图（侯仁之主编：《北京历史地图集·政区城市卷》，文津出版社）
Sketch map of Dadu of the Yuan Dynasty during the Zhizheng era (*The Historical Atlas of Beijing · Administrative Regions and Cities* edited by Hou Renzhi, Wenjin Publishing House)

gods, an inheritance of the tradition of setting sacrificial places in the southern suburb.

In 1553, Emperor Jiajing constructed the Outer City, using the perimeter of city walls to safeguard and manage the prosperous residential and commercial area ouside Zhengyangmen Gate, as well as the Altar of Heaven and the Earth and the Altar of Mountains and Rivers. In 1553, the Yongdingmen Gate as the main south entrance to the Outer City was completed, which was a contribution to the grand scale present-day Beijing Central Axis that totals 7.8 kilometers.

Five pavilions enshrining five statues of Budda were constructed at the top of the Jingshan Hill during the reign of

明初在南京营建过程中形成的一套皇宫布局形态，被完整地迁移到北京。由于明代北京城是在元大都的基础上进行了「北缩南扩」，原本位于城市南部的中轴线，在新的城市建设过程中就成了贯穿整个城市南北的北京中轴线。

The complete layout of royal palaces that developed during the construction of the capital Nanjing was then fully applied to the contruction of Beijing. As the Ming Dynasty capital was relocated southward from the site of Dadu of the Yuan Dynasty, the original central axis in the city's southern part became the Beijing Central Axis that ran through the entire city.

符合儒家秩序观念的理想中的新都城。大都城的中轴线是城市的核心，呈现出与从曹魏邺城到隋唐长安城的大朝殿与南向皇家大道不同的特征。大都城开启了一个中国都城的新形态。

明初定都南京，元大都降格为北平府，大内宫殿遭拆毁，北部空阔的区域被放弃，城市的范围有所缩减。明永乐皇帝决定从南京迁都北京后，1407年开始兴建北京内城及宫城、皇城。明初在南京营建过程中形成的一套皇宫布局形态也被完整地迁移到了北京。由于明代北京城在元大都的基础上进行了"北缩南扩"，原本位于城市南部的中轴线，在新的城市建设过程中就成为贯穿整个城市南北的北京中轴线。明代北京中轴线在延续元代"面朝后市"，将对城市进行管理的钟楼和鼓楼置于轴线北端的同时，在皇城南部紧邻中轴线的东西两侧设置了太庙和社稷坛，使中轴线的布局更为紧凑、严谨。在中轴线向南延长线的两侧布置了天地坛和山川坛，祭祀天地山川诸神，这也是南郊祭祀传统的延续。

1553年明嘉靖皇帝加筑北京外城，将正阳门外繁华的居住区、商业区及天地坛、山川坛通过修建外城城墙保卫和管理起来。同年外城正南门永定门建成，最终使北京中轴线形成了今天长达7.8公里的规模。

1750年清乾隆皇帝增建了景山山脊上的五座亭子，并供奉五座佛像，在景山北侧中轴线上移建寿皇殿建筑群，使北京中轴线的建筑形态得到进一步加强。

1912年清王朝宣布宣统皇帝逊位之后，中国从封建王朝迈入现代国家。北京从帝王的都城转化为市民的城市，北京中轴线上原来帝王的宫殿、坛庙相继成为博物馆、市民公园。天安门前的宫廷广场、千步廊等先后被拆除。1949年天安门广场成为新中国举行开国大典的地方，成为新中国首都的中心和国家政治与礼仪活动的中心。北京中轴线在中华人民共和国的建设过程中，依然得到了尊重与延续。1958年人民英雄纪念碑在北京中轴线上、天安门广场中心落成。1959年天安门广场两侧分别建成中国革命博物馆及中国历史博物馆、作为人民参政议政和国家重大国事活动场所的人民大会堂。两组建筑以北京中轴线为中心对称布局，东侧的中国革命博物馆及中国历史博物馆与太庙隔长安街南北相对，西侧的人民大会堂与社稷坛隔长安街南北相对。1976年毛泽东主席逝世，1977年毛主席纪念堂在北京中轴线上、人民英雄纪念碑南侧、正阳门北侧落成。天安门广场上的这些建筑是20世纪下半叶中国最为重要的建筑，它们在北京中轴线上的建设不仅是北京中轴线物质形态的发展和延伸，更是北京中轴线所表达的传统精神的延续。

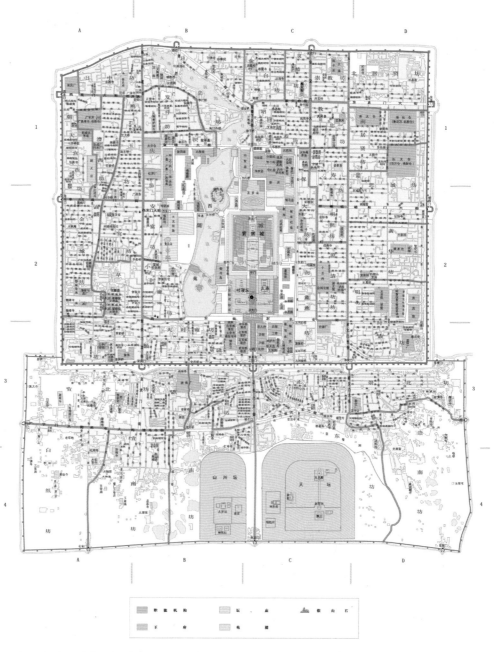

万历至崇祯年间的明北京城示意图 (侯仁之主编:《北京历史地图集·政区城市卷》,文津出版社)
Sketch map of the city of Beijing from Wanli to Chongzhen reigns in the Ming Dynasty (*Historical Atlas of Beijing · Administrative Regions and Cities* edited by Hou Renzhi, Wenjin Publishing House)

Emperor Qianlong in the Qing Dynasty. The architectural complex of the Shouhuang Hall was relocated to the north of the Jingshan Hill, further reinforcing the architectural form of Beijing Central Axis.

Following the abdication of Emperor Xuantong in 1912, China began to embrace a modern era. Beijing transformed itself from a royal capital for monarchs to a city for the people. Emperors' palaces, temples and altars on the central axis were successively turned into museums or parks. The court square and the one-thousand-step corridor in front of the Tian'anmen Gate Tower were demolished. In 1949, Tian'anmen Square saw the founding ceremony of the People's Republic of China, becoming the center of the new capital and the political and ceremonial center of the state. Beijing Central Axis was inherited and respected in the construction process of Beijing. In 1958, the Monument to the People's Heroes was completed, located at the center of Tian'anmen Square on Beijing Central Axis. In 1959,

明代北京中轴线在延续元代『面朝后市』，将对城市进行管理的钟楼和鼓楼置于轴线北端的同时，在皇城南部紧邻中轴线的东西两侧设置了太庙和社稷坛，使中轴线的布局更为紧凑、严谨。在中轴线向南延长线的两侧布置了天地坛和山川坛，祭祀天地山川诸神，这也是南郊祭祀传统的延续。

While following the layout of "the court in the front and the marketplace in the back" and setting the Bell and Drum Towers as city management facilities at the north end of the axis line, city planners of the Ming Dynasty arranged the Imperial Ancestral Temple and the Altar of Land and Grain on east and west sides of the central axis respectively, close to the southern section of the Imperial City, giving rise to a more compact and stricter layout of the central axis. On both sides of the southern extension of the central axis were the Altar of Heaven and the Earth and the Altar of Mountains and Rivers to offer sacrifices to various gods, an inheritance of the tradition of setting sacrificial places in the southern suburb.

新中国成立后基于首都建设的需求，大部分的城墙和城门被拆除。20世纪80年代以后随着历史文化名城保护观念的发展，人们开始重新关注北京中轴线及相关遗产的保护。2005年永定门城楼被重建，成为北京中轴线南部端点的地标。北京中轴线也呈现出今天古今交融的独特形态。

北京中轴线的文化意义与遗产价值

北京中轴线的价值体现在多个方面：

北京中轴线反映了中国传统的"以中为尊"观念。

北京中轴线不仅是城市空间的规划和组织，更是一种传统观念的表达，"以中为尊"是这种观念最为重要的组成部分。中国古代文献中记载，"古之王者，择天下之中而立国，择国之中而立宫，择宫之中而立庙"，"谓之地中，天地之所合也，四时之所交也，风雨之所会也，阴阳之所和也。然则百物阜安，乃建王国焉"。作为中国传统价值观的核心，在元大都城规划之初，无论是筑中心台确定城市中心，还是将最重要的建筑群组置于中轴线上，都是这种观念在北京城市物质形态上的反映。元代之后，明、清两代对北京中轴线的扩建和完善本身均反映了这种观念的延伸。20世纪50年代选择在天安门广场建筑人民英雄纪念碑以及70年代在天安门广场建设毛主席纪念堂，也都表达了对这种传统观念的尊重和延续。

北京中轴线上各组建筑、建筑群的对称与对位关系同样反映了对中正和秩序的追求。这种观念不仅仅影响到建筑的布局或北京中轴线的体形形态，更重要的是这种物质形态本身是中国传统观念的自觉表达，这种观念已经成为中国人道德意识的组成部分。

北京中轴线是多个历史时期的积淀，有丰富的历史文化遗存。

北京中轴线源于元代大都城的规划建设。万宁桥作为北京中轴线的遗产构成要素，是元代大都城中轴线位置的标志，也是元代的文物遗存。万宁桥也反映了当时京杭运河的漕运在积水潭及周边地区形成繁荣的商业活动，是"面朝后市"城市格局的重要见证。万宁桥自身同样具有很高的艺术价值。桥拱上的兽头、桥下东西两对镇水兽，体现出当时城市建设过程中的审美追求，表现了艺术与功能的融合。故宫在近年的考古发掘中也发现了元代地层，为认识元代大内的状况提供了研究材料。

北京中轴线保留了许多明代的重要建筑遗存，其中鼓楼、太庙、社稷坛、先农坛的主要建筑

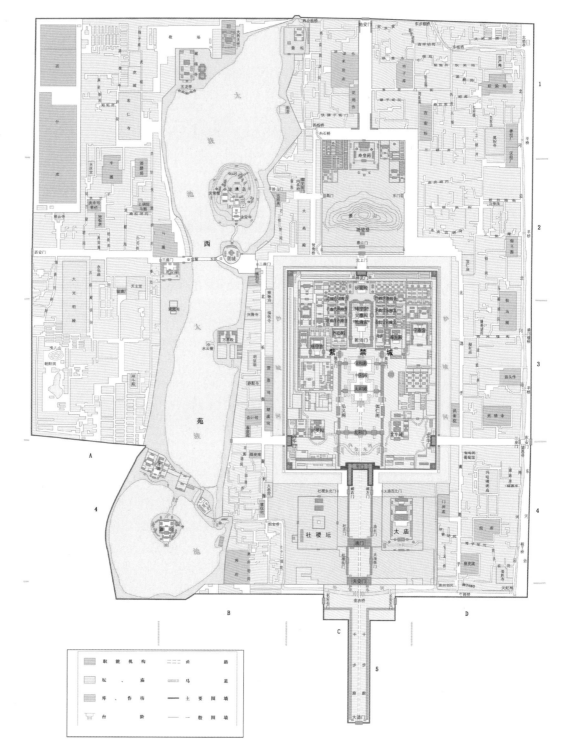

乾隆十五年 (1750年) 的清皇城示意图 (侯仁之主编：《北京历史地图集·政区城市卷》, 文津出版社)
Sketch map of Qing Dynasty Imperial City in the 15th Year of Qianlong Reign (*Historical Atlas of Beijing · Administrative Regions and Cities* edited by Hou Renzhi, Wenjin Publishing House)

construction work was completed for the Museum of Chinese Revolution and the Museum of Chinese History on the east side of Tian'anmen Square, and the Great Hall of the People on its west side as a venue for people's participation in deliberation and administration of state affairs and for hosting significant state ceremonies and activities. These two groups of buildings are symmetrically arranged, with Beijing Central Axis at the core: the Museum of Chinese Revolution and the Museum of Chinese History face the Imperial Ancestral Temple on the north side of Chang'an Avenue, while the Great Hall of the People faces the Altar of Land and Grain on the north side of Chang'an Avenue. In

为明代遗存。太庙前殿(享殿)的规模、等级与太和殿和明长陵祾恩殿相同,是中国现存规模最大的木结构殿堂建筑。

钟楼、景山、正阳门城楼、天坛的主要建筑是清代在明代基础上重建、改建和增建的部分。其中景山山脊上以位于北京中轴线上的万春亭为中心、东西对称布置的五座亭子是清乾隆十五年(1750年)建造的,这是清代对北京中轴线的一次重要完善。乾隆皇帝在景山五亭中供奉佛像,也赋予了北京中轴线更为丰富、复杂的文化内涵。

清代末年,铁路等交通设施进入北京,部分城墙被打通或拆除,并对原有的城楼等建筑进行改建,以适应当时的审美趣味。正阳门箭楼1916年完成立面的装饰工程,反映了这一时期北京城市建设的特征。

1949年中华人民共和国成立,20世纪50年代和70年代对天安门广场的改建延续了北京中轴线的重要地位。人民英雄纪念碑、中国国家博物馆(原中国革命博物馆和中国历史博物馆)、人民大会堂、毛主席纪念堂等重要纪念建筑和公共建筑的建设延续了北京中轴线反映的传统规划观念,这些建筑也表现了这一时期的艺术风格。

北京中轴线是中国传统理想都城模型的重要展现。

元初忽必烈、刘秉忠规划大都城时,对《周礼·考工记》国都形态的附会是对传统理想都城模式的实践。《周礼·考工记》关于都城形态的描述,是对基于等级、礼仪制度的城市秩序的追求,是中国文化传统中,通过建立秩序形成人与环境之间的积极关系的实践。北京中轴线则是这一实践的结晶。无论是"面朝后市,左祖右社"的布局形态,还是尊卑有序,张弛有度的城市空间组织,都表达了对这种秩序的追求。明代嘉靖皇帝对于相关礼仪、秩序的修正和确认,也反映了这种都城理想模式的意义。北京中轴线上的相关遗存是这种中国传统理想都城模式实践的见证。

北京中轴线承载了中国复杂的祭祀礼仪,见证了业已消逝的城市管理方式。

由于大量坛庙建筑的存在,北京中轴线也是多样复杂、等级繁复的传统祭祀礼仪的载体。它承载了不同时间、不同形态的祭祀活动,反映了中国传统文化中"国之大事,在祀与戎"的传统。

城市,特别是都城的秩序,不仅体现在功能布局和空间形态上,同样也体现在城市中人的行为模式上。在古代城市中,通过各种管理机构、设施、方式对城市中人的行为进行管理,城门和市场的开闭、里坊制度、宵禁等都是中国古代城市管理的重要方面。北京融合了自秦统一之后的都城管理的各种方式。北京中轴线上遗存的城门是中国古代多重城墙制度的印证。永定门是北京外城最为重要的中央城门,正阳门和正阳门箭楼是北京内城的正门,天安门和现

民国时期的北京城中轴线俯瞰图(侯仁之主编:《北京历史地图集·政区城市卷》,文津出版社)
A bird's eye view of Beijing Central Axis during the Republican period (*Historical Atlas of Beijing · Administrative Regions and Cities* edited by Hou Renzhi, Wenjin Publishing House)

北京中轴线上不同时代的历史遗存
和建筑见证了近八个世纪中华文明
的发展，以及中轴线所反映的中国
传统文化观念的延续。

Historical remains and buildings from
different eras that stand along Beijing
Central Axis bear testimony to the
development of Chinese civilization in
nearly eight centuries and the continuation
of Chinese traditional culture.

1976, Chairman Mao passed away. In 1977, the Chairman Mao Memorial Hall was constructed on Beijing Central Axis, located on the south side of the Monument to the People's Heroes and the north side of the Zhengyangmen Gate. The buildings at Tian'anmen Square are the most important monuments built in the second half of the 20th century. They embody not only the development and extension of the physical form of Beijing Central Axis but also the continuation of the spirit it conveys.

For the needs of the development, most city walls and gates in Beijing were demolished. From the 1980s, as conservation ideas for historical and cultural cities developed, people began to renew their attention to the protection of Beijing Central Axis and its related heritage properties. In 2005, the Yongdingmen Gate Tower was reconstructed which has become a landmark at the southern end of Beijing Central Axis. Today, Beijing Central Axis presents a unique form that connects the past with the present.

Cultural significance and heritage value of Beijing Central Axis

The value of Beijing Central Axis is manifested in several aspects as follows:

Beijing Central Axis reflects the idea of "respect for the central" in Chinese traditions.

Symmetric and corresponding relationships among different buildings and architectural ensembles along Beijing Central Axis also reflect a pursuit of centrality and order in the traditional culture. This conception has influenced the architectural layout and physical form of Beijing Central Axis, and more importantly, this physical form itself is an expression of traditional ideas that comprise the moral system of the Chinese people.

Beijing Central Axis embodies the inheritance from several historical periods and contains a great wealth of historical and cultural remains.

Historical remains and buildings from different eras that stand along Beijing Central Axis bear testimony to the development of Chinese civilization in nearly eight centuries and the continuation of Chinese traditional culture.

Beijing Central Axis is an important presentation that displays an ideal capital city from Chinese traditions.

The form of capital city, as described in the *Artificer's Record* in the *Rites of Zhou*, conveys the pursuit of city order based on hierarchical and ritual systems and crystalizes traditional cultural practices to develop an interactive relationship between man and nature (heaven) through the creation of order. Beijing Central Axis, in terms of either its layout featuring "the court in the front and the marketplace in the back, the ancestral temple on the left and the altar of land and grain on the right" or its organization of urban spaces following the principle of hierarchy and moderation, manifests such a pursuit of order. Historical remains related to Beijing Central Axis bear witness to such practices to build an ideal capital city from Chinese traditions.

Beijing Central Axis carries China's sophisticated sacrificial rites and ceremonies and bears witness to a city management approach that has already disappeared.

With numerous temples and altars, Beijing Central Axis serves as a carrier for diverse, sophisticated and hierarchical sacrificial rites and ceremonies from different periods and in different forms. It reflects the Chinese tradition that "sacrificial rites and military affairs are the state's top priorities".

The order of cities, especially capital cities, is presented not only by their functional layouts and spatial forms but also by the behavioral model of their residents. In ancient times, citizens' behaviors were regulated through various

钟鼓楼(袁雪飞／摄)
Bell and Drum Towers (Photo by Yuan Xuefei)

已不存的地安门是北京皇城的城门，午门和神武门则是宫城的城门，它们反映了中国都城的圈层管理、防御特征及这种独特的城市结构。

钟楼和鼓楼是中国古代城市时间管理的机构和公共建筑，通过报时的方式实现对城市活动的管理。无论是官员的早朝，还是城门和市场的开闭，城市生活的方方面面都是通过钟楼、鼓楼的报时来管理的。钟楼和鼓楼是中轴线上最为高大的建筑，并处于北京中轴线的北端，这也显现了其重要性。

北京中轴线上不同时代的历史遗存和建筑见证了近八个世纪中华文明的发展，以及中轴线所反映的中国传统文化观念的延续。

北京中轴线是独特的都城规划方法的见证。

元大都的规划与建设是对中国古代理想城市形态的重现。在继承传统城市规划的基础上，元大都的规划者创造了新的规划方法：筑台确定城市中心点，确定城市规模和四至范围，确定城市核心轴线方位。这种规划方法以新的方式确定了城市的功能分布和"面朝后市，左祖右社"等级清晰、井然有序的城市形态。北京中轴线是这种新的城市规划方法的见证。这种规划方法得到了明、清两代的尊重和延续，也深刻地影响了北京当代城市的发展。

北京中轴线构成了北京核心独特的连续、富有韵律的建筑组群。

不同于曹魏邺城到隋唐长安形成的以大朝正殿为起点、以大朝正殿前皇家大道为轴线的组织城市功能、确定城市形态的规划方法，元代以中心台为起点确定城市中轴线并在中轴线上设置重要建筑群的规划方法，使北京中轴线上形成了多重功能复杂的建筑群，并随着明、清两代及近现代的进一步加强和丰富，这一特征变得更为突出。

北京中轴线的建筑空间可以划分为几个大段落：

钟楼、鼓楼到万宁桥，是北京自元大都时期以来的繁华商业区。积水潭、什刹海的云影波光倒映着钟楼、鼓楼高耸华丽的身影，与繁华的街巷、商铺构成了一幅热烈、丰富的市井景象。

management services, facilities and methods. The opening and closing of city gates and markets, the residential area system, and imposition of curfews were important elements of city management in ancient China. Beijing has inherited various methods of capital city management developed following the unification by the Qin Empire. Surviving city gates on Beijing Central Axis are the evidence of city wall defensing system in capital cities of ancient China. The Yongdingmen Gate is the important main gate to the outer city; the Zhengyangmen Gate and its Archery Tower constitute the main entrance to the inner city; the Tian'anmen Gate and the once standing Di'anmen Gate served as gates to the imperial city; and the Meridian Gate and the Gate of Divine Prowess were gates to the palace city. These gates all manifest the unique urban structure of capital cities in ancient China that features multiple circles of urban management and defense.

The Bell and Drum Towers are time-keeping institutions and public buildings for cities in ancient China. All aspects of city life were regulated through the announcement of time from the Bell and Drum Towers, from holding the court to the opening and closing of city gates and markets. The Bell and Drum Towers are the tallest buildings on Beijing Central Axis, standing at its northern end to highlight their important status.

Beijing Central Axis bears testimony to a unique method for capital city planning.

Dadu of the Yuan Dynasty is representation of an ideal city form from ancient China. By following traditional city planning practice, its planners created a new planning method: fixing the central point of the city by building a central platform, determining the scale of the city and its boundaries on four sides, and locating the core axis of the city. This new planning method defines how city functions are arranged and gives rise to a hierarchical, well-ordered city form that features "the court in the front and the marketplace in the back, the ancestral temple on the left and the altar of land and grain on the right". Beijing Central Axis is the evidence to testify to this new city planning method which has continued to be respected and adopted in the ensuing Ming and Qing dynasties and has had profound influence on the city's development in modern times.

Beijing Central Axis constitutes unique continuous and rhythmic architectural ensembles at the core of the city.

Architectural spaces of Beijing Central Axis can be roughly divided into several major sections as follows:

The area ranging from the Bell and Drum Towers to Wanning Bridge has been the city's booming commercial area ever since the Yuan Dynasty. The shimmering Jishuitan and Shichahai lakes and the lofty Bell and Drum Towers are combined with bustling streets and shops to form dynamic, diversified scenes of urban life. The Jade River encircling the area and Wanning Bridge as its pause point present a space that resembles a natural scenery surrounded by a mountain and girdled by a river. The "marketplace at the back" lies against the Bell and Drum Towers, with the Jade River in the front and the Jingshan Hill on the opposite, establishing itself not only as a commercial area in the north section of Beijing Central Axis but also as an independent, perfect area. The Jingshan Hill to the north of the Forbidden City and the Inner Jinshui River and the Inner Jinshui Bridge in front of the Gate of Supreme Harmony are

从故宫遥望景山（李少白／摄）
Jingshan Hill as seen from the Forbidden City (Photo by Li Shaobai)

北京中轴线将皇家祭祀空间、外城商业空间、官廷广场（现代政治中心）、皇家宫室、御苑、皇城、内城市场等建筑群体布局在中轴线上，形成富于变化、连续而活跃的空间序列，成为北京不同于其他都市的城市空间形态。

Beijing Central Axis contains a cluster of architectural ensembles, including imperial sacrificial area, Outer City commercial area, imperial court square (modern political center), royal palaces and gardens, the Imperial City, and Inner City marketplaces. All these are combined to form a varying, continuous and dynamic spatial order that distinguishes Beijing from any other capital city.

玉河对这一区域的环绕与万宁桥的收束呈现出类似于山环水抱的空间特征。这一"后市"区域背倚钟楼、鼓楼，前临玉河，对望景山，既是整个北京中轴线北端的商业空间，又自成一体，构成一个完美的空间环境。景山作为紫禁城宫殿的靠山与太和门前的内金水河及跨越其上的内金水桥，同样形成了一个完整的空间环境。午门到天安门前的外金水河及外金水桥使这样的空间形态再次呈现。天安门、天安门广场与正阳门及正阳门外的内护城河和正阳门桥，又是类似的空间形态。从正阳门到天桥，由天桥及两侧《正阳桥疏渠记》碑，《帝都篇》《皇都篇》碑两座碑亭构成的双阙与桥的形态组合标记了这一空间范围的边界。从天桥到永定门，以两侧的天坛和先农坛为核心的皇家祭祀区域则以相对自然的环境特征表达了对古代郊祀习俗的记忆。

北京中轴线将这样一组皇家祭祀空间、外城商业空间、宫廷广场（现代政治中心）、皇家宫室、御苑、皇城、内城市场等建筑群体布局在中轴线上，形成富于变化、连续而活跃的空间序列，成为北京不同于其他都市的城市空间形态。从这些连续的组合空间的形态特征可以看到，其每一个组成部分又都呈现出传统理想空间的特征，这些空间组合在一起则构成了反映古代理想都城的城市形态。

北京中轴线构成了独特的城市历史景观。

北京中轴线上连续、富有韵律的建筑群集中了元、明、清及现当代中国最重要的纪念性建筑和等级最高的建筑，它们高耸于北京老城中心，以巨大的体量、绚丽的色彩和丰富的建筑造型成为独特的城市历史景观。这一城市历史景观对周边环境产生了明显的影响，也形成了今天北京历史文化名城保护的区划和控制条件要求的基础。

从最重要的建筑和建筑群组之间关系的层面，各建筑之间采用了严格的对称和对位关系，形成了连续、严谨的空间和景观联系。中国传统官式建筑在形式上的一致性进一步加强了这一建筑群体的整体特征。出于对传统布局形式、建筑风格的尊重，在20世纪后半叶对天安门广场的改、扩建中，新的建筑与原有中轴线建筑组群的形态高度协调，融为一体。

在空间形态上，北京中轴线从北向南，重要建筑、空间、桥梁水系的反复穿插与多重组合形成了独具特色的都城空间形态，这本身也构成了严谨秩序之下又富于变化、活跃的城市景观。

北京中轴线是从13世纪到20世纪中国反映东方哲学和世界观的都城形态和空间秩序的典范。

combined to create, once again, an integrated layout of space. The area from the Meridian Gate to the Outer Jinshui River and the Outer Jinshui Bridge in front of the Tian'anmen Gate is another demonstration of such spatial form. The Tian'anmen Gate and Tian'anmen Square, combined with the Zhengyangmen Gate, the Inner Moat, and the Zhengyang Bridge, present yet another example of such spatial form. In the area from the Zhengyangmen Gate to Tianqiao (heavenly bridge), Tianqiao and two stele pavilions together form a combined view of twin Que towers and a bridge, marking boundaries of the area. The imperial sacrificial area from Tianqiao to the Yongdingmen Gate, with the Temple of Heaven and the Altar of Agriculture on both sides of the central axis, conveys memories of ancient suburban sacrificial traditions through features of natural environment.

Beijing Central Axis contains a cluster of architectural ensembles, including imperial sacrificial area, Outer City commercial area, imperial court square (modern political center), royal palaces and gardens, the Imperial City, and Inner City marketplaces. All these are combined to form a varying, continuous and dynamic spatial order that distinguishes Beijing from any other capital city.

Beijing Central Axis presents a unique historical urban landscape.

Continuous, rhythmic architectural ensembles along Beijing Central Axis bring together the most important monuments and highest-grade buildings from the Yuan, Ming and Qing dynasties as well as modern China. Standing at the center of Beijing's old city, architectural ensembles of Beijing Central Axis present a unique historical urban landscape with their huge sizes, bright colors and diverse designs. This landscape has remarkable influence over the surrounding environment and provides a basis for zoning and control for the conservation of Beijing as a historical and cultural city.

Beijing Central Axis is a paradigm of urban layout and spatial order of capital cities evolving in China from the 13th to 20th century which is a reflection of Oriental philosophy and worldview.

Beijing Central Axis manifests the traditional Chinese belief system that is diverse and sophisticated.

The idea of "all have a spirit" enjoys a profound social foundation in China, a traditional agricultural nation. This belief is also expressed in Beijing Central Axis. The construction of important state temples and altars and arrangement of various sacrificial rites clearly demonstrate the diversity and richness of the traditional Chinese belief system, including the worship of nature, Confucianism, Buddhism and Taoism.

Beijing Central Axis bears forms of everyday life from different classes of Chinese society.

As an architectual ensemble and a city axis that runs across Beijing's old city, Beijing Central Axis clearly

乾清门（范炳远／摄）
Gate of Heavenly Purity (Photo by Fan Bingyuan)

太庙（阮旭红／摄）
Imperial Ancestral Temple (Photo by Ruan Xuhong)

北京中轴线反映了中国传统多样复杂的信仰体系。

中国作为一个传统的农业国家，万物有灵的思想具有深厚的社会基础，这种信仰关系在北京中轴线上也同样得到了表达。

祖先崇拜是中国信仰体系的重要组成部分，在北京中轴线上存在多处与祖先崇拜相关的遗存。太庙是其中最为重要的一处，其建筑形制的变迁，从合祀制度到分祀制度，再到合祀制度的变化，反映了祖先崇拜在中国文化传统中的重要地位。故宫内的奉先殿是帝王在宫内告祭祖先的地方。景山北侧的寿皇殿建筑群是乾隆十五年（1750年）移建至北京中轴线上的供奉已故帝后画像的地方。

社稷坛是祭祀土地神和五谷神的地方，象征生存空间和粮食生产，这二者的结合则成为国家和政权的象征。社稷坛五色土筑的祭坛顶面形象地表达了国土的概念。"左祖右社"表达了中国传统社会中最重要的信仰内容和它们之间的关系，反映了中国文化中"家国一体"的政治观念。

明代在规划北京时，在南郊北京中轴线延长线的两侧设置了天地坛和山川坛，供奉天地山川的各路神仙。明代嘉靖改制（1530年）后，天地坛改为天坛，成为之后明、清两代帝王祭祀皇天上帝和行祈谷礼的地方。山川坛则改为先农坛，祭祀先农（农神）、太岁（年神）。清顺治时又在外坛南部增建天神坛和地祇坛。

这些重要国家坛庙的设置以及各种祭祀活动的安排清晰地反映了中国传统的信仰体系。

除了这些反映中国泛神信仰的坛庙之外，在故宫中轴线的最后一座殿堂钦安殿中供奉了道教中真武大帝的坐像，真武大帝是道教中的北方之神，明成祖朱棣认为他能够得到皇位是获得了真武大帝的帮助，因此他不仅规划修建了真武大帝的圣山武当山的宫观道场，也在钦安殿供奉了真武大帝神像，这一供奉形式

社稷坛五色土（金东俊／摄）
Five colors of earth in Altar of Land and Grain (Photo by Jin Dongjun)

displays forms of everyday life from different classes of Chinese society.

Beijing Central Axis runs northward through the Outer City, the Inner City, the Imperial City, the Palace City and commoners' living areas in Beijing's old city. Among others, the section from Tianqiao to the Zhengyangmen Gate in the Outer City is the residential area dwelled by commoners. This area sees the living state of commoners in ancient China and enjoys the greatest wealth of intangible cultural heritage in Beijing. The Inner City section along Beijing Central Axis, such as the area from the Di'anmen Gate to the

天安门 (fotoe.com／供图)
Tian'anmen Gate (Provided by fotoe.com)

Bell and Drum Towers as well as its surrounding historical quarters, maintains many princely mansions and noblemen's residences, reflecting the living state of the upper class, in particular the nobilities. The Forbidden City showcases life of emperors from the Ming and Qing dynasties. Today, Beijing Central Axis presents life of contemporary people in Beijing and transformation of the city from an imperial capital to a new capital city for the people.

Beijing Central Axis witnesses many significant historical events.

As the center of East Asian civilizations, China has undergone numerous events of world infulence. From the rise and fall of the Yuan Dynasty to the changes of reigns in the Ming Dynasty, from the establishment of the Qing Dynasty to the abdication of its last emperor, Beijing Central Axis has seen all those events, and has turned to be their material witness.

In 1949, Tian'anmen Square has witnessed the founding ceremony of the People's Republic of China when Chairman Mao Zedong, standing on the terrace of the Tian'anmen Gate Tower, solemnly proclaimed its founding. This, among other significant events, has exerted great impact on the world.

Since the 13th century, Beijing Central Axis has been an important carrier of significant events, belief systems and social lives of worldwide impact.

遥相呼应 (王心超／摄)
A Dialogue between the Historical Architectures (Photo by Wang Xinchao)

北京中轴线从南向北穿越了北京老城的外城、内城、皇城、宫城和内城的市井区域。其中外城从天桥到正阳门区段是平民百姓的居住区域，这一区域也是北京非物质文化遗产资源最为丰厚的地区。

Beijing Central Axis runs northward through the Outer City, the Imperial City, the Palace City and commoners' living areas in Beijing's old city. Among others, the section from Tianqiao to the Zhengyangmen Gate in the Outer City is the residential area dwelled by commoners. This area sees the living state of commoners in ancient China and enjoys the greatest wealth of intangible cultural heritage in Beijing.

在清代也得到了延续。此外，清代乾隆时期还在景山南麓北京中轴线上建绮望楼，楼中供奉孔子，供官学学生祭拜，并在景山山脊建五亭，供奉佛像。

除这些占据重要位置并用作供奉场所的坛庙、建筑之外，在北京中轴线的许多建筑或空间中还有庙宇、神龛，供奉各种神明，其中包括正阳门瓮城中的关帝庙和观音寺、故宫周围的诸多庙宇、万宁桥西侧的火德真君庙等。北京中轴线展现了中国传统信仰体系的多样性和复杂性。

北京中轴线是中国社会各阶层生活形态的载体。

作为一条纵贯北京老城的城市轴线和建筑空间群体，它清晰地展现了位于这条轴线上的中国各个社会阶层的生活形态。

北京中轴线从南向北穿越了北京老城的外城、内城、皇城、宫城和内城的市井区域。其中外城从天桥到正阳门区段是平民百姓的居住区域，这里反映了中国古代社会普通民众的生活状态，这一区域也是北京非物质文化遗产资源最为丰厚的地区。清代初期将北京老城的内城划为八旗驻地，将原居住在城内的汉族居民、官员全部迁入外城居住。北京中轴线的内城部分，如地安门到钟鼓楼及周边的历史街区，保留了大量反映王公贵族及社会上层生活形态的王府和贵族住宅。故宫则反映了明、清两代帝王的生活。今天的北京中轴线又呈现了北京从帝王都城到人民城市和共和国首都的当代生活。

北京中轴线是许多重要历史事件的发生地。

北京中轴线与许多具有世界性影响的重大事件有着直接的关联。中国作为东亚文明的中心，无论元朝的建立、灭亡，明朝的更迭，清王朝的兴起与1912年清王朝宣告皇帝退位，都具有地区性和世界性的影响。北京中轴线是这些事件的发生地，是这些事件的物质见证。

1949年10月1日开国大典在天安门广场举行，毛泽东主席在天安门城楼庄严宣布中华人民共和国成立。北京中轴线也是许多重要历史事件的发生地。

北京中轴线是13世纪以来中国具有世界影响的重大事件、信仰体系和社会生活的重要载体。

北京中轴线的独特性

北京中轴线是独特的。这种独特性在于它本身是中国传统文化、传统世界观的反映，是中国传统文化具有代表性的载体。北京中轴线是中国文化观念的表达，是一种中国气派。

中国传统文化，特别是以儒家学说为代表的传统文化，希望通过建立一种秩序来促使人类

前门大街（袁雪飞／摄）
Qianmen Street (Photo by Yuan Xuefei)

The unique nature of Beijing Central Axis

Beijing Central Axis is unique as it manifests Chinese traditional culture and worldview. Beijing Central Axis is a physical and representative expression of Chinese traditional culture and its grandeur.

The capital city construction system from the *Artificer's Record* in the *Rites of Zhou* has a longstanding influence and became the ideal model when Dadu of the Yuan Dynasty was planned. It proved the immense influence of Confucianism over the Chinese society.

"Respect for the central" is a Chinese traditional idea. Beijing Central Axis determines the symmetrical layout of the whole city, which is the expression of the principle on the urban scale.

Beijing Central Axis presents the ideal order of Chinese capital cities through the spatial layout featuring "the court in the front and the marketplace at the back, the ancestral temple on the left and the altar of land and grain on the right" and the symmetrical urban layout. The immense scale of architectural ensembles along Beijing Central Axis plays a governing role in developing and regulating the city's order.

Chinese civilization has continued uninterrupted for thousands of years. Over the historical development of world cities, most cities would give rise to different planning ideas in different periods, which influenced transformation of the cities. Beijing Central Axis expresses the core spirit of Chinese traditional culture which has continued along with the development of the central axis itself. The planning philosophy not only governed the form of Dadu of the Yuan Dynasty when it was firsted constructed in the 13th century, but also determined the reconstruction when the Ming Dynasty relocated its capital to Beijing in the 15th century, as well as the addition of the Outer City in the 16th century. It is also this planning philosophy that influenced and determined the characteristics of the city of Beijing in the 20th century, and will further shape the future of Beijing in the 21st century and beyond. This is a rare phenomenon in the human history of city development. It manifests not only features of living heritage at Beijing Central Axis but also the continuation and transmission of Chinese traditional culture.

社会和谐运行。这种建立严谨秩序的愿望在儒家学说中以各种方式表达出来，无论是家族血缘的父子关系，还是国家统治的君臣关系，都强调克制个人的欲望，遵循礼仪秩序，成就和谐的社会关系和秩序。西周的分封制度是中国传统等级制度和礼制关系的起始，也是儒家思想的重要基础。《周礼·考工记》的营国制度是基于西周分封制度的城市等级体系，这一体系与儒家的礼制、秩序是一致的。这种营国制度能够具有长久影响且在元代规划新首都大都城时成为附会的对象及理想都城模型，表现了儒家学说在中国社会的巨大影响力。忽必烈在大都城的规划中运用这样的规则，也反映了儒家思想对东亚地区的广泛影响。

"以中为尊"是中国传统观念，而皇权更强调了"以中为尊"对于国土进行有效管治的要求。清代乾隆皇帝在《帝都篇》中曾对历代都城做了评述，他认为北京作为都城，西有太行，东临沧海，南襟河济，北据居庸，既方便海上交通，又得运河之利，具有其他古都无法比拟的重要地位。在都城中，在中轴线上设置宫殿、坛庙就是对"以中为尊"思想的表述。

北京中轴线对中国都城理想秩序的表达不仅体现在传统的"面朝后市，左祖右社"的空间形态和"九经九纬，经涂九轨"的对称城市格局，同样也体现在北京中轴线建筑群巨大体量对整个城市秩序的统领作用。在城市建筑的形态上，北京中轴线上的建筑是北京城内最高等级的建筑，它们以高大的体量影响和控制着周边的建筑和景观环境。在城市空间上，北京中轴线处于城市的核心位置，东西两侧的街巷以北京中轴线为轴对称布局，内城的街巷呈现出方格网及道路正交的形态。甚至重要建筑的名称也是以北京中轴线为核心对称命名的，如对称于永定门东西两侧的外城南城墙的两个城门分别称为左安门、右安门；以正阳门为中，左右对称设置在内城南城墙的两个城门则分别称为崇文门和宣武门。这种秩序还伴随着城市中的各种行为规范，包括通过钟鼓楼的报时对城内生活的管理。通过北京中轴线建立井然有序的城市，反映了中国文化观念对秩序的追求。

按城市职能划分，世界历史上的城市大体可以分为两类，一类是作为统治中心的政治权力城市，一类是作为工商业发展中心的商业城市。北京则是历史上政治权力城市的代表。在这类城市中往往需要通过规划和建筑的意向表达权力的至高力量，北京中轴线正是这种意向的表达。在世界城市发展进程中，大部分城市在不同发展时期都会呈现出不同的规划思想，影响城市形态的变化。而北京在始建时就以中心轴线控制整个城市的布局，并在之后各个扩建过程中始终通过这一中心轴线的扩展控制城市形态的变化，这一规划建设在世界都城中极具独特性。

中华文明历经五千年绵延不断。北京中轴线表达了中国传统文化的精神核心，这种规划思想伴随着北京中轴线的延续而延续。它不仅决定了13世纪元大都初建时的形态，也决定了15世纪初明代迁都改建北京、16世纪北京扩建外城时的城市形态，同样影响和决定了20世纪北京的城市特征和21世纪城市发展的未来。这在人类城市发展历史上是罕见的现象，不仅反映了北京中轴线活态遗产的特征，也体现了中国传统文化观念的延续与传承。

北京中轴线申报世界遗产的意义

2009年北京中轴线申报列入《世界遗产名录》工作启动，2013年被列入中国申报世界遗产的《预备清单》。2017年伴随全国文化中心建设，北京中轴线申报世界遗产工作全面提速展开。

北京中轴线的保护和传承在历次北京总体规划中都被反复提出和强调。随着价值研究、保护和环境整治的开展，对北京中轴线的遗产价值及其对当代社会发展的影响、对延续近八个

皇穹宇的圆形围垣——回音壁（董亚力／摄）
The Echo Wall Enclosing the Imperial Vault of Heaven (Photo by Dong Yali)

Relevance of World Heritage nomination for Beijing Central Axis

Beijing Central Axis began its World Heritage nomination process in 2009 and was included in the Tentative List in 2013. With the advancing of the construction of Beijing as a national cultural center in 2017, the nomination work sped up in full swing.

The conservation and continuation of Beijing Central Axis has been reiterated through the drafting and revisions of the master plan of Beijing. With academic research of the values of Beijing Central Axis and its conservation and environmental rehabilitation, significant improvements have been made with regard to its heritage value, its influence over the development of contemporary society, and the recognition of its physical features as the most integrated Chinese capital city for nearly eight centuries. A clear image of Beijing Central Axis has gradually unrolled.

The World Heritage nomination of Beijing Central Axis has relevance in several aspects as follows:

It is a process of reviewing the value of related heritages from an integral perspective, telling stories about Chinese history and culture, and interpreting the unique and continuous nature of Chinese civilization. The value identification of Beijing Central Axis is a process to shift focus from its physical form to the culture and spirit it embodies; from its influence over city planning and development in different periods to cultural relevance and spiritual connotations it expresses, such as "respect for the central", creation of ritual order, and pursuit of an ideal capital city model; from its ancient architectural complexes to its surviving living heritages; from city planning to historical urban landscape. This extended and deepened understanding of the value and cultural connotations of Beijing Central Axis has given rise to more narrative perspectives

地祇坛九座山岳、江海纹石龛（冬青、紫叶小檗组成花坛，模拟地祇坛坛台）（欧阳平／摄）
Nine stone niches with mountain, river and sea design in the Altar of Earthly Gods (Photo by Ouyang Ping)

北京中轴线表达了中国传统文化的精神核心，这种规划思想伴随着北京中轴线的延续而延续。它不仅决定了13世纪元大都初建时的形态，也决定了15世纪初明代迁都改建北京、16世纪北京扩建外城时的城市形态，同样影响和决定了20世纪北京的城市特征和21世纪城市发展的未来。这在人类城市发展历史上是罕见的现象……

Beijing Central Axis expresses the core spirit of Chinese traditional culture which has continued along with the development of the central axis itself. The planning philosophy not only governed the form of Dadu of the Yuan Dynasty when it was firsted constructed in the 13th century, but also determined the reconstruction when the Ming Dynasty relocated its capital to Beijing in the 15th century, as well as the addition of the Outer City in the 16th century. It is also this planning philosophy that influenced and determined the characteristics of the city of Beijing in the 20th century, and will further shape the future of the city of Beijing in the 21st century and beyond. This is a rare phenomenon in the human history of city development.

世纪保存最为完整的中国都城形态特征的认识都有了巨大提升，北京中轴线的形象也越来越清晰地展现出来。

北京中轴线申报世界遗产工作具有以下意义：

整体审视相关遗产的价值，讲好中国历史文化故事，讲好中国文明的独特性和延续性。北京中轴线的价值梳理经历了一个从物质形态到文化精神的认识过程：从开始关注北京中轴线对不同时期城市规划和发展的影响，进一步发展到其所表达的"以中为尊"、建立礼制秩序、追求理想都城模式等文化意义和精神内涵；从古代建筑群到延续至今的活态遗产；从城市规划到城市历史景观的不断发展。这种对北京中轴线价值和文化内涵认识的扩展和深化，形成了更多讲述中国历史文化遗产故事的角度和方式。2018年、2019年北京市政府先后组织了两届关于北京中轴线的国际研讨会，2020年在新冠肺炎疫情影响下又召开了线上研讨会。在这三次研讨会上，随着对北京中轴线价值研究的展开，也从不同方面阐释了北京中轴线的遗产价值，讨论了申报世界遗产的方法和可能的途径。

申报世界遗产也是一个保护和展现的过程。在历史发展进程中，由于功能的改变和城市性质的变化，北京中轴线的各个构成要素都或多或少存在不利于表达其历史文化价值的问题。申遗过程的保护和整治，就是探寻北京中轴线作为一个整体如何展现它最为理想的形态，展现北京中轴线连续、深厚的历史文化内涵，展现中国首都延续750余年的中国气派。

作为一项仍然保持着旺盛生命力的历史遗产，北京中轴线不仅属于创造它的历史年代，同样也是当代生活的载体，影响着人们对今天北京的认知和理解，更影响着未来北京的城市发展。北京中轴线这一北京老城的历史轴线，随着对其价值和意义认识的深化，它越来越强烈地影响北京中轴线南、北延长线的规划和建设。它将更强烈地促进北京城市的古今融合，促进北京未来的城市发展。

申报世界遗产是中国作为《保护世界文化和自然遗产公约》缔约国的国际责任，我们有义务将反映中国文化精神和中国创造性智慧、反映中华民族历史和文化多样性、反映中华民族历史文化发展的遗产展现在世界面前，并保护好它们，使之传之久远。

认识这些遗产、保护好这些遗产也是传承中华民族优秀文化传统的必要条件。保护、展示好这些遗产，形成良好的城市形态和城市景观，使今天的人们能够感受到遗产所内含的文化精神，能够树立文化自信心和自豪感，能够更好地在传承基础上创新发展。

and approaches to present and interpret Beijing Central Axis as a historical and cultural heritage property which has Outstanding Universal Value.

The World Heritage nomination is a process of protecting and presenting the heritage property. As city functions and natures change the sweep of history, components of Beijing Central Axis have been more or less occupied or altered by functions that have adverse effect on the expression of its historical and cultural value. The conservation and rehabilitation work carried out in the nomination process is intended to explore how Beijing Central Axis as an integrated whole can present its ideal form, its continuous and profound historical and cultural connotations, and its grandeur as a capital city of China that survives over 750 years.

Beijing Central Axis is a historical heritage property that is still full of vigor and vitality. It not only belongs to the time when it was created, it also serves as a carrier to sustain contemporary life, influencing people's understanding of Beijing today and the development of Beijing in the future. Along with ever-deepening recognition of its value and relevance, Beijing Central Axis, as a historical axis in the old city, has ever-increasing influence on the planning, extension and construction of its southern and northern sections and will more remarkably promote the integration of the past with the present and the future development of Beijing.

World Heritage nomination is an international obligation of China as a State Party to the *Convention Concerning the Protection of the World Cultural and Natural Heritage*. It is our obligation to show the world those heritage properties that reflect the spirit of Chinese culture, creative wisdom of Chinese people, and historical development and cultural diversity of Chinese nation, and to take good care of them so as to have them carried long onward.

Understanding and taking good care of these heritage properties are necessary conditions for transmitting Chinese cultural traditions. Protecting and presenting these heritage properties and developing urban forms and landscapes can enable contemporary people to recognize the cultural spirit embodied by them, foster a sense of pride and confidence in our culture, and make innovations on the basis of transmitting traditions.

日落时分，中轴之美（朱雨生／摄）
View of the Central Axis at sunset (Photo by Zhu Yusheng)

The Sundown Drum and the Daybreak Bell

钟鼓楼—地安门段落

Bell and Drum Towers – Di'anmen Section

—— 钟鼓楼 ———————————————— 万宁桥 ——

地安门外大街 ——————— 地安门

钟楼南立面（金东俊／摄）
The southern facade of the Bell Tower (Photo by Jin Dongjun)

夜幕下的钟鼓楼（王心超／摄）
The Bell and Drum Towers at night (Photo by Wang Xinchao)

钟鼓楼—地安门段落位于地安门以北、什刹海东岸,自元代起就是商贾云集的"后市"所在,也是北京民俗文化最为活跃的地带,这里见证了北京中轴线最初的生成过程。

钟鼓楼坐落在北京中轴线北段,钟楼是中轴线的北端点。钟鼓楼曾是古都北京的报时设施,中国农耕社会长期形成了"暮鼓晨钟"的报时文化,反映了古人"日出而作,日入而息"的传统生活方式,古人用这种声音传播方式建立起时间秩序乃至社会的秩序。

在元大都时期,开启了都城"暮鼓晨钟"的报时传统。鼓楼上置鼓,钟楼内悬钟,鼓楼击鼓定更,钟楼撞钟报时,周而复始,这一传统一直延续到明清时期。元明清三代,钟鼓楼都是北京城重要的报时中心之一,文武百官上朝和市民的劳作生息都以钟鼓声为准。钟鼓楼还是皇权的象征,其报时功能对社会秩序的维系,本身就具有强烈的政治意义。

随着时代变迁和科技的进步,钟鼓楼在近现代不再承担报时功能,而逐渐具备了服务公众的教育、休闲与游览等作用,时至今日成为市民休闲放松的好去处。

从元代建造之初开始,万宁桥就一直担负着重要的交通作用。从陆路上看,它与地安门外大街共同构成中轴线在该地区的南北交通要道;从水路上看,万宁桥起到了将元代大运河通惠河段与今什刹海(元积水潭)沟通的作用。这座桥梁更为广阔的意义是,从元代以来万宁桥的建筑形制虽略有变化,但其位置与作用始终不变,因此佐证了元大都及明清北京中轴线位置叠压关系,证明了北京中轴线有着悠久的历史。

地安门内、外大街向北通达鼓楼、钟楼,向南延伸到景山。长期以来,地安门外大街市井商业繁荣,而明清时期处于皇城内的地安门内大街则显得庄重且静谧。

The section from the Bell and Drum Towers to Di'anmen Gate lies north of Di'anmen Gate and east of Shichahai Lake. It has become the "marketplace at the back" since the Yuan Dynasty, witnessing bustling commercial and trade activities and most dynamic folklore practices and giving rise to the early development of Beijing Central Axis.

The Bell and Drum Towers stand at the northern section of Beijing Central Axis with the Bell Tower marking its north end. They were timekeeping facilities for the administration of the imperial capital in ancient times. "Daybreak bell and sundown drum" was a longstanding tradition in China's agricultural society, symbolizing the lifestyle of ancients who "get up at sunrise and work until sunset". This unique way of sound transmission brought an order of time and even an order of social governance to the imperial capital of Beijing.

The timekeeping tradition of "daybreak bell and sundown drum" was originated from the period of Dadu of the Yuan Dynasty. The bell was striken and the drum was beaten to tell time all day long and all year round. This practice continued throughout the Ming and Qing dynasties. With the changes of the times and the progress of technology, the Bell and Drum Towers no longer serve as timekeeping facilities in modern times; instead, they function as an educational, leisure and sightseeing place today.

Wanning Bridge has long been an important transportation facility since its construction in the Yuan Dynasty. It is not only an overland facility that combined with Di'anmen Outer Street, forms a north-south avenue along the central axis, but also a water transportation facility that connects the Tonghui Section of the Grand Canal of the Yuan Dynasty with the present-day Shichahai Lake (Jishuitan Lake in the Yuan Dynasty).

Di'anmen Outer and Inner Streets start from the Bell and Drum Towers in the north and extend to Jingshan Hill in the south. Over centuries, Di'anmen Out Street has been a bustling commercial area for commoners while Di'anmen Inner Street, located within the Imperial City during the Ming and Qing dynasties, looks solemn and quiet.

钟鼓楼
Bell and Drum Towers

民国时期的北京鼓楼(老明信片)
The Drum Tower in the Republican era (In an old postcard)

钟鼓楼作为明清两代的城市管理设施屹立数百年,是中国古代都市独特的时间管理方式和公共生活方式的见证。

高耸的钟楼与魁梧的鼓楼一北一南,两座建筑之间由一道狭长的广场连接,共同构成了简单而实用的公共空间。

The Bell and Drum Towers, functioning as city management facilities for several hundred years during the Ming and Qing dynasties, bear witness to the unique time management method and way of public life practiced in an imperial capital of ancient China.

20世纪80年代后期，鼓楼、钟楼向广大公众开放，成为公众参观古迹、登高远眺的景点。

In the late 1980s, the Bell and Drum Towers were opened to the public as museums, where visitors can appreciate historic monuments and enjoy a distant view.

从西北方向看钟楼（绿汀／摄）
Bell Tower as seen from northwest (Photo by Lyu Ting)

钟鼓楼高耸于居民区中（朱雨生／摄）
The Bell and Drum Towers overlooking low-rise courtyard houses (Photo by Zhu Yusheng)

从钟楼上南望鼓楼（牛飞／摄）
The Drum Tower viewed from the Bell Tower in the north (Photo by Niu Fei)

钟楼
The Bell Tower

在北京中轴线的钟鼓楼段落上，
钟楼居北端，是一座重檐砖石建筑，
建筑外有菱形围墙环绕，占地面积5700多平方米。

The Bell Tower marks the north end of Beijing Central Axis.
It is a double-eave masonry structure encircled by diamond-shaped walls.
The whole building compound occupies an area of 5,700 square meters.

钟楼南立面（金东俊／摄）
The southern facade of the Bell Tower (Photo by Jin Dongjun)

暮鼓晨钟 钟鼓楼—地安门段落

从鼓楼西大街看钟楼（王华龙／摄）
View of the Bell Tower from the West Gulou (Drum Tower) Street (Photo by Wang Hualong)

古代钟鼓楼如何报时？
How the Bell and Drum Towers told time in the past?

元大都时期，钟楼内放钟，鼓楼内设漏壶、鼓角，报时系统由此建立。至明清时期，这一报时方式得到了延续。明代钟楼最初使用铁钟报时，后来觉得声音不够洪亮，改换铜钟，鼓楼上放置了铜刻漏、漏壶和铙等计时、报时设备，铜刻漏上标有刻度，每过一刻度，人工击铙八下，进行报时。到了清代，报时系统更加完善，钟楼在白天正午会报时，在夜间五个更次也会报时，鼓楼则改用香和鼓来报时，用燃香来定更次，定时击鼓。

During the period of Dadu of the Yuan Dynasty, the bell was installed in the Bell Tower and the water clock and drums and horns in the Drum Tower to put in place a time service system. This way of telling time continued during the Ming and Qing dynasties. The Ming Dynasty first used an iron bell to tell the time, but later it was replaced by a bronze bell in order to amplify the sound. During the Qing Dynasty, the Bell Tower would chime at noon and at the five watches at night; while for the Drum Tower, incents were burned to determine the night watches, and the drum was beaten accordingly to tell time.

钟鼓楼街区的市民生活 (左图牛飞／摄，右图闫立军／摄)
Folklife near the Bell and Drum Towers (Left photo by Niu Fei, right photo by Yan Lijun)

钟楼内的明代大钟 (金东俊／摄)
The giant bell from Ming Dynasty in the Bell Tower (Photo by Jin Dongjun)

钟鼓楼街区保留着自古以来的市井活力（金东俊／摄）
Lively public space at the foot of the Bell and Drum Towers (Photo by Jin Dongjun)

鼓楼
The Drum Tower

位于钟楼之南,是一座木结构重檐歇山三滴水建筑,占地面积6800多平方米,外有围墙环绕。
钟鼓楼之间的广场,向北至钟楼南围墙,向南接鼓楼北门,呈长方形,
南北长约85米,东西宽约30米,目前是市民活动的空间。

Located south of the Bell Tower,
the Drum Tower is a wood structure of a gable and hip roof with doube eaves and triple layers of drip tiles encompassed by walls.
The rectangular square between the Bell Tower and the Drum Tower,
about 85-meter long from south to north,
and 30-meter wide from west to east, is now a public space for citizens.

鼓楼南立面(金东俊／摄)
The southern facade of the Drum Tower (Photo by Jin Dongjun)

今天钟鼓楼广场是市民活动的好去处（金东俊／摄）
The square of the Bell and Drum Towers is a popular public space (Photo by Jin Dongjun)

鼓楼内的击鼓表演 (陆岗／摄)
The drum-beating performance in the Drum Tower (Photo by Lu Gang)

钟鼓楼鸣响"北京时间"
The "Beijing Time" as is defined by the Bell and Drum Towers

钟鼓楼是明清时代北京城的报时中心，均由钦天监管理。其中，钟楼撞钟报时辰，于每日黄昏鸣钟，随后起更，次日清晨再鸣钟一次；鼓楼击鼓定更次，于每日交更(晚7时)击鼓，随后每更次击鼓一通。文武百官听到三通鼓便起床，四通鼓时赶至紫禁城午门，五通鼓则鱼贯入朝。这种报时是当时北京的标准时间，也是朝廷每天授时的重要标志。清代由銮仪卫署派专人敲钟、击鼓，指导着百官上朝及京城千家万户的起居劳作。

鼓楼内陡直的阶梯 (牛飞／摄)
Steep staircases in the Drum Tower (Photo by Niu Fei)

The Bell and Drum Towers were the city's timekeeping center during the Ming and Qing period. The Bell was chimed at dusk and again in the early morning of the next day, while the Drum was beaten once every two hours from seven o'clock in the evening onward. The sound of the Bell and the Drum indicated standard time to guide court attendance for officials and everyday life for commoners in the imperial capital.

万宁桥
Wanning Bridge

万宁桥桥身微拱，部分望柱与栏板为历史原构(郝毅／摄)
Wanning Bridge is a slightly arched structure with some of its balusters and balustrades being the original components from ancient times (Photo by Hao Yi)

万宁桥始建于元代至元二十二年(1285年)，最初是一座木构桥，后改为石砌桥，现存为单拱石桥,桥面宽约17米,长约35米。元代初建时,万宁桥是大都城内通惠河上游的重要通水孔道,横跨在什刹海入玉河口处,桥下西侧设置闸门,当时叫海子闸,后改名为澄清闸,万宁桥因此兼具桥与闸的双重功能。

First constructed in 1285 in the Yuan Dynasty, Wanning Bridge was originally a wooden structure and later converted into a stone structure. The current bridge is a single-arch stone bridge, 17 meters in width and 35 meters in length. During the Yuan Dynasty, it provided an important water channel at the upper stream of the Tonghui Canal in Dadu, functioning as both a bridge and a sluice.

万宁桥是世界遗产大运河北京段的组成部分（绿汀／摄）
Wanning Bridge is a heritage component of the Beijing Section of the Grand Canal World Heritage (Photo by Lyu Ting)

万宁桥边设有元明时期的镇水兽和水闸（牛飞／摄）
Close to Wanning Bridge are the stone beast guards and a sluice from the Yuan and Ming dynasties (Photo by Niu Fei)

镇水石兽
River Guardian Beasts

2000年，万宁桥在修缮过程中清理出几只被埋多年的镇水石兽，经考证，是元、明时期的遗存。这些石兽是传说中的龙子蚣蝮，能吞江吐雨，调节水量，所以常设在桥洞或建筑的排水口，以避水患。

The River Guardian Beasts found during the repair work of Wanning Bridge are remains from the Yuan and Ming dynasties. They are not only ornamental components but also benchmarks to measure the depth of the canal.

桥西侧北岸的镇水兽(郝毅／摄)
The River Guardian Beast on the north bank west of the bridge (Photo by Hao Yi)

桥东侧北岸的镇水兽为元代遗存(绿汀／摄)
The River Guardian Beast on the north bank east of the bridge is a relic of the Yuan Dynasty (Photo by Lyu Ting)

桥西侧南岸的镇水兽(范炳远／摄)
The River Guardian Beast on the south bank west of the bridge (Photo by Fan Bingyuan)

镇水兽伏岸观水，有镇水保平安之意(范炳远／摄)
The River Guardian Beast at the canal's shore for peace and prosperity (Photo by Fan Bingyuan)

万宁桥西侧景观(王飞／摄)
Scenery west of the Wanning Bridge (Photo by Wang Fei)

万宁桥西与什刹海相连(陆建成／摄)
Shichahai Lake to its west (Photo by Lu Jiancheng)

位于万宁桥边的火神庙在明清时期曾香火兴旺(绿汀／摄)
The Temple of Fire God next to the Wanning Bridge used to be busy in the Ming and Qing dynasties (Photo by Lyu Ting)

地安门外大街
Di'anmen Outer Street

修葺中的鼓楼与地安门街景（李彦成主编：《中轴旧影》，文物出版社）
Old photo of the Drum Tower under restoration and Di'anmen Street (*The Old Photos of the Central Axis of Beijing* edited by Li Yancheng, Cultural Relics Publishing House)

地安门外大街在元代即为著名的商业街道。元代营城时根据"前朝后市"的思想,把主要市场集中于皇宫大内与中心台之间的区域,京杭大运河的通惠河开通,积水潭成为繁忙的漕运码头。于是自元代起,这里沿街店铺林立,商贾云集,商品琳琅满目,集市人声鼎沸,如《析津志》所言,"本朝富庶殷实,莫盛于此"。地安门外大街延续了商贸区的功能,直至今日。

Di'anmen Outer Street became a popular commercial street from the Yuan Dynasty, bringing together numerous shops and merchants. During the Ming and Qing period, while the area inside the Di'anmen Gate was a forbidden quarter, the area outside continued to function as a commercial district.

城

万宁桥以北至鼓楼的地安门外大街（袁雪飞／摄）
Di'anmen Outer Street connecting the Drum Tower to the north and Wanning Bridge to the south (Photo by Yuan Xuefei)

俯拍地安门外大街 (牛飞／摄)
Aerial view of Di'anmen Outer Street (Photo by Niu Fei)

地安门外大街上有烟袋斜街的入口,是元代时就沿积水潭北岸形成的商业街(牛飞／摄)
The entrance to a commercial street from the Yuan Dynasty in Di'anmen Outer Street
(Photo by Niu Fei)

什刹海
Shichahai Lake

万宁桥以西的什刹海本是一个狭长湖泊，早年水域广阔，元代紧邻湖泊东岸确定了大都城规划设计的北京中轴线，并为明清北京城沿用。以什刹海、太液池等水草丰美的水域为核心来确定城市格局，体现了元代草原游牧文化与中原农耕文化的融合。

由于漕运兴盛和风光优美，这一街区不仅商业繁荣，更吸引了历朝历代的官宦贵胄在此设置王府、园林，加之寺庙庵观的大量兴建，这里还成为古代京城的文化与民俗重地。

Shichahai Lake was originally a narrow, long lake. Because of prosperous commerce in the neighborhood, princely mansions and gardens and Buddhist and Taoist temples were constructed here over past dynasties, making it a place rich in culture and folklore in ancient times.

左上角为万宁桥（龙脉航拍／供图）
Wannning Bridge is visible on the upper left (Provided by Longmai Aerial Image)

繁华依旧的古老街区（陆岩／摄）
Old street of lasting vitality (Photo by Lu Yan)

什刹海夏日（陆岗／摄）
Summer at the Shichahai Lake (Photo by Lu Gang)

人力车载着游客走街串巷(陆岩／摄)
Rickshaws for tourists (Photo by Lu Yan)

什刹海老宅门前的糖葫芦(陆岗／摄)
Sugar coated haws in front of an old house in Shichahai area (Photo by Lu Gang)

层叠的瓦片屋顶上留下雨雪和岁月的印记（陆岩／摄）
Traditional roof tiles of courtyard houses in winter time (Photo by Lu Yan)

从什刹海看向钟鼓楼（王心超／摄）
The Bell and Drum Towers viewed from Shichahai Lake (Photo by Wang Xinchao)

地安门 ——————— 地安门内大街 ——————— 景山

皇城宫苑

Imperial Palaces and Gardens

地安门—天安门段落
Di'anmen Gate – Tian'anmen Gate Section

故宫 ——————————————— 太庙
社稷坛 ——————————————— 天安门

景山万春亭与牌坊（刘雯／摄）
Wanchun Pavilion and the archway at Jingshan Hill (Photo by Liu Wen)

从地安门到天安门的皇家宫苑区,为古代的皇家文化提供了见证实例。

位于中轴线核心位置的故宫是明清两代皇帝举行朝政、皇室起居生活的空间,是古代社会里皇权统治的至高象征。故宫以北的景山是明清时期皇家苑囿的重要组成,是皇城中重要的观景高地。位于故宫东南角的太庙是皇帝祭祀祖先的家庙,故宫西南角的社稷坛则是祭祀土地和五谷之神的场所。从分布上看,太庙与社稷坛紧邻北京中轴线,呈对称布局,恰好反映了《周礼·考工记》中"左祖右社"的格局,是古代都城理想范式应用于北京中轴线规划格局的直接体现。

地安门内大街在元代为皇家御苑,在明清时期为皇家衙署所在,之后逐渐向公众开放。

景山是明清两代皇家御苑,古人巧妙地堆叠山形地势,形成独具特色的园林景观。作为北京中轴线上的制高点,景山提供的视觉景观将中轴线南北紧密地联系起来,从而营造出南北延展的如画江山。景山建筑群不只是景观,它还具有祭祀先祖的重要作用,为明清两代奉行的国家礼仪祭祀制度与传统提供了见证。

故宫旧称"紫禁城",始建于明永乐四年(1406年),它既是永乐大帝以来的明代、清代的宫城所在,也是北京中轴线的核心。整组宫殿建筑群布局严谨有序,规划布局与建筑形制严格遵守了中国传统文化中的礼仪秩序规定,体现出鲜明的以中轴线左右对称的格局,以此展现出皇帝至高无上的威仪。故宫占地面积约72万平方米,建筑面积约15万平方米,是目前世界上规模最大、保存最为完整的木结构古建筑群之一。作为中国古代宫殿建筑的杰出作品,故宫集大成地展现了中国古代王朝的宫廷文化与皇家礼仪特点。

太庙是明清两代皇家的祖庙,用来祭祀历代祖先。它是明清两代北京老城内最重要的皇家礼制建筑之一,也是中国目前保存最完整、规模最大的皇家祭祖建筑群。作为皇室祭奠祖先的场所,太庙象征着王朝的统治权在本家族内代代传递的合法性。在太庙进行的祭祖活动,是中国传统礼制活动中的最高等级,具有至高无上的地位。

社稷坛是明清两朝皇家祭祀农耕之神——社神、稷神的最高等级坛庙,也是国家政权的象征之一,希冀疆土永固、物产丰富。从1914年开始,社稷坛作为城市公园对公众开放,成为北京老城内皇家坛庙建筑转变为公园的第一例,开启了北京中轴线向近现代公众空间转变的过程。

The imperial palace and garden complex stretching from Di'anmen Gate to Tian'anmen Gate provides an example that bears physical witness to the royal culture in the past.

Jingshan Hill was an imperial garden during the Ming and Qing dynasties. The garden's fanciful arrangement of artificial hills and terrains shaped its unique landscape. It was also a place for royal families of the Ming and Qing dynasties to worship their ancestors.

Formerly known as the "Forbidden City" (today's Palace Museum), the Imperial Palace had been the Palace City of the Ming and Qing dynasties since Ming's Yongle reign, and it was also at the core of Beijing Central Axis. The layout of the complex of palaces was well planned in rigorous order. Layout and form of the architectures closely followed Chinese tradition of ritual order and rules. The clear symmetrical layout along the central axis shows the supreme majesty of the emperor.

The Imperial Ancestral Temple was for ancestor worshiping of the royal family in the Ming and Qing dynasties, where sacrificial ceremonies were held. It was one of the most important ritual architectures in the old city of Beijing during the Ming and Qing dynasties. It is also the best preserved and the largest surviving imperial ancestral temple complex. The Altar of Land and Grain is the highest-ranking architecture of its kind where Ming and Qing emperors worshiped those agricultural deities as the gods of land and grain. As one of the symbols of state power, it also epresses the royal family's aspiration for a forever blessed territory and abundant products.

地安门内大街
Di'anmen Inner Street

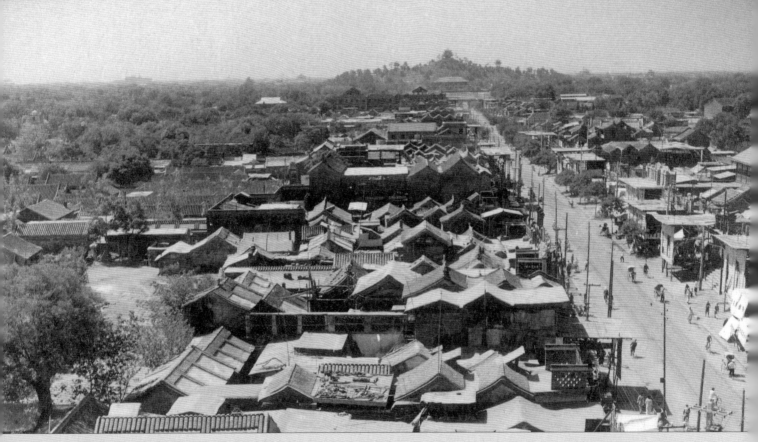

从地安门大街至景山旧貌（李彦成主编：《中轴旧影》，文物出版社）
Old photo showing the view of Di'anmen Street leading to Jingshan Hill in the south (*The Old Photos of the Central Axis of Beijing* edited by Li Yancheng, Cultural Relics Publishing House)

　　地安门内大街位于北京中轴线中部北段，向北与地安门外大街相连，向南对接景山北门，全长约550米，宽约45米。在明清两代，地安门内大街是皇城北门向南通往景山的道路，两侧聚集了一些大内衙署、库局。进入民国后，大街向公众开放，逐渐成为市民居住街区。

　　Di'anmen Inner Street is located in the north of the middle section of Beijing Central Axis. It connects Di'anmen Outer Street in the north and the north entrance to Jingshan Park in the south, totaling 550 meters in length and 45 meters in width. In the Ming and Qing dynasties, Di'anmen Inner Street started from the north gate to the Imperial City and stretched toward Jingshan Hill, flanked by imperial government offices and storage for the royal court. In the Republican period, the street was put to public use and gradually turned into a residential area for commoners.

通往景山的地安门内大街(牛飞/摄)
Di'anmen Inner Street leading to Jingshan Hill (Photo by Niu Fei)

景山
Jingshan Hill

1892年，北京景山 (fotoe.com／供图)
Jingshan Hill in 1892 (Provided by fotoe.com)

　　景山坐落于明清的宫城即紫禁城的北面,是北京中轴线上具有承前启后作用的景观节点和全城的制高点。作为皇家御苑,景山与紫禁城共同位于都城中轴线之上,南北相邻,这一独特布局方式延续了古代中国都城的长期传统,诸如北魏洛阳城、隋唐长安城等,体现了用皇家御苑拱卫宫城的意图。

　　Being a garden for the royal family, Jingshan Hill was the highest point that links the views on Beijing Central Axis. Jingshan Hill and the Forbidden City sit next to each other on the central axis of the capital. This unique arrangement is a continuation of the long tradition of a capital city from ancient China that royal gardens function as a barrier for the imperial palaces.

南望景山 (牛飞／摄)
Jingshan Hill as seen from the north (Photo by Niu Fei)

四柱三间九楼式的牌坊 (牛飞／摄)
A traditional archway (Photo by Niu Fei)

北京城南北中轴线标识 (阮旭红／摄)
Marker of North-South Central Axis of Beijing (Photo by Ruan Xuhong)

　　景山地处明清北京内城的中心区域，山体最高峰距地面高度约为45.7米，是中轴线上的最高点，南邻庄严肃穆的紫禁城，北望位于闹市的钟鼓楼。作为明清时期的皇家御苑，景山给皇家提供了登高望远以及休憩的功能；在清代，这里还增添了部分祭祀功能。

　　Jingshan Hill is located in the center of the Ming and Qing dynasties' Inner City of Beijing. The highest peak is about 45.7 meters from the ground, the highest point on the central axis. Sitting in its south is the dignified Forbidden City, while to its north are the Bell and Drum Towers amid the hustle and bustle of the marketplace.

绮望楼
Qiwang Tower

绮望楼是供奉孔子牌位的建筑，
它依景山山脚而修建，建筑面阔五间，进深三间，重楼重檐形制，明间挂着满汉文书写的匾额"绮望楼"；
前带廊，四周设有汉白玉石护栏。
Qiwang Tower is where the spirit tablet of Confucius is enshrined. Built at the foot of the Jingshan Hill,
it is a two-story double-eave structure of a width of five bays and a depth of three bays with white marble balustrades built around it.

从绮望楼仰望景山万春亭 (金东俊／摄)
Qiwang Tower and Wanchun Pavilion in its rear (Photo by Jin Dongjun)

万春亭望远（金东俊／摄）
A distant view from the Wanchun Pavilion (Photo by Jin Dongjun)

　　景山作为人工堆叠的土丘，其主峰正位于北京内城南北中点，选址与地形塑造明显含有中国传统的厌胜意味，故景山又被称为"镇山"，以祈求皇权永续。山形高耸峻拔，树木翁郁，以此延续中国古代御苑祭祀自然神灵的礼仪传统。《明宫史》记载，天子每逢重阳日都登万岁山远眺全城，至清代于景山上设五亭，这里更成为北京老城内登高远眺的最佳视点。景山以视觉联系北京中轴线南北：向南俯瞰紫禁城，向北眺望钟鼓楼。

　　Jingshan Hill is an artificial mound built in the Ming Dynasty on the ruins of the palace of the Yuan Dynasty. Its main peak stands at the mid-point of the Inner City's north-south axis. The chosen site and terrain shaping were obviously meant to fend off the enemies, a tradition in ancient China. Because of this, it is also called "Zhen Shan", or hill of suppression, in order to pray for everlasing reign of the emperor.

景山五亭
Five Pavilions of Jingshan Hill

景山上有五亭，
正中主峰为万春亭，其余四亭两两左右对称，高度逐次降低，
东有观妙亭、周赏亭，西为辑芳亭、富览亭，东西两侧亭台在平面形制、尺寸、屋顶形态与色彩等方面均严格对称。

There are five pavilions on Jingshan Hill.
At the main peak in the center is Wanchun Pavilion and the other four pavilions are located in balanced symmetry,
with their heights gradually reduced.
The pavilions on the east and west sides are arranged in strict symmetry with regard to layout, dimension, style of the roof and colors.

景山之夜（陆岗／摄）
Jingshan Hill at night (Photo by Lu Gang)

观妙亭 (绿汀／摄)
Guanmiao Pavilion (Photo by Lyu Ting)

富览亭 (顾彩华／摄)
Fulan Pavilion (Photo by Gu Caihua)

从故宫东北角楼处看景山 (王飞／摄)
Jingshan Hill as seen from the northeast Corner Tower of the Forbidden City
(Photo by Wang Fei)

　　乾隆十五年 (1750年) 在景山上以中轴线为轴对称建造了五亭，成为北京中轴线上重要的观景制高点。清代对景山建筑格局的改造，丰富了北京中轴线的空间序列，加强了其对称格局。

　　In 1750, Qing Emperor Qianlong built the five pavilions, a high vantage point on the Beijing Central Axis offering wonderful view of the palace.

寿皇殿建筑群
Shouhuang Hall Building Complex

景山内的寿皇殿建筑群原本位于山体东北侧，
清乾隆十四年(1749年)由原来的位置移建于山体正北面，
从而加强了中轴线上的对称布局，并扩大了原有建筑群规模，里面供奉清代帝后们的御像。

In 1749, Qing Emperor Qianlong relocated Shouhuang Hall to due north of Jingshan Hill and expanded the scale of the complex.
The expansion was to make room for paying homage to Qing emperors and empresses.

寿皇殿建筑群是中轴线上除故宫外的第二大建筑群(顾彩华／摄)
Shouhuang Hall building complex is the second largest architectural ensemble on the Central Axis, after the Forbidden City (Photo by Gu Caihua)

渣土堆变成了景山
A mound of excavated earth turned into Jingshan Hill

　　北京城内本无山。永乐年间营建紫禁城时，开修护城河挖掘出的泥土将其堆积成山，并在山上广植树木，形成景山，成为明清两代的皇家御苑和故宫风水学意义上的"靠山"，也曾称为镇山、煤山。

　　Beijing has no mountain within the city. During the construction of the Forbidden City in the Yongle era, more than a million cubes of earth excavated in forming the moats were piled up into a hill - Jingshan Hill, and trees were planted. It served as the backing hill to the imperial palaces in terms of Geomantic principles.

寿皇殿是中国古代高等级的建筑形制 (牛飞／摄)
Shouhuang Hall is a high-grade architecture in ancient China (Photo by Niu Fei)

寿皇殿内部 (牛飞／摄)
Interior of the Shouhuang Hall (Photo by Niu Fei)

寿皇殿内部 (金东俊／摄)
Interior of Shouhuang Hall (Photo by Jin Dongjun)

　　在景山的中轴线上，寿皇殿坐落于北端。它坐北朝南，建筑形制等级极高，面阔九间，进深五间，前檐出廊，采用重檐庑殿顶形制；龙头须弥石座台基一重，汉白玉望柱栏板围绕，月台深远开阔，上面放置了四座铜鼎炉以及铜鹤、铜鹿等。

　　Shouhuang Hall is on the northern tip of the central axis of the Jingshan Hill. Sitting in the north facing south, it was in an architectural style of the highest ranking.

故宫
Forbidden City

《平定准部回部得胜图·平定回部献俘》中的午门献俘场面,清代郎世宁绘
Presenting Prisoners to the Emperor, by Giuseppe Castiglione, Qing Dynasty

　　故宫北以景山为靠，南邻天安门。故宫在北京中轴线上的位置关系生动地展现出中国古代世界观的想象图景，也展示出古人对理想社会模式的想象。故宫始建于明永乐四年(1406年)，为明清两代宫城。故宫的整组宫殿建筑群布局秩序井然,高低宽窄、一砖一瓦以及装饰色彩等都严格遵循礼仪秩序,并在整体上鲜明地体现出居中和左右对称的布局。建筑群的宏大布局,凸显帝王至高无上的威严。

　　The Forbidden City is at the heart of Beijing Central Axis, with Jingshan Hill to its north and the Tian'anmen Gate in the south. Built in the fourth year (1406) of the reign of the Ming Emperor Yongle, the Imperial Palace was the Palace City during the Ming and Qing dynasties. Palace buildings in the cluster are arranged in strict sequence. The height and width of the buildings, the position of bricks and tiles, even the decorations and colors all closely adhere to the ritual order. The entire arrangement is a vivid demonstration of the layout plan of a central location and strict symmetry.

端门
Upright Gate (Duanmen)

端门位于午门和天安门之间，
整个建筑结构和风格与天安门相似。
在明清两代，端门的城楼主要是存放皇帝仪仗用品的地方。

The Upright Gate is located between the Meridian Gate (Wumen) and Tian'anmen Gate.
Its architectural structure and style resemble those of the Tian'anmen Gate.
During the Ming and Qing dynasties,
its gate tower served as storage of the emperor's ceremonial objects.

雪漫故宫，前景为端门一角，后景是午门（王飞／摄）
Snowfall in the Imperial Palace, with one corner of the Upright Gate in the foreground and Meridian Gate at the back (Photo by Wang Fei)

端门前的日晷（张月军／摄）
Sundial in front of the Upright Gate (Photo by Zhang Yuejun)

明清皇宫：中国第一批世界遗产
Imperial Palace of the Ming and Qing Dynasties: the First Batch of World Heritage in China

1987年，明清皇宫与周口店北京人遗址、泰山、长城、秦始皇陵及兵马俑坑、莫高窟一同入选《世界遗产名录》，成为中国第一批世界遗产。

In 1987, the Imperial Palace of the Ming and Qing Dynasties, together with the Peking Man Site at Zhoukoudian, Mount Taishan, the Great Wall, Mausoleum of the First Qin Emperor and the pit of the Terracotta Army, and the Mogao Caves, were inscribed on the *World Heritage List*, the first batch of World Heritage in China.

前朝区
Front Court

故宫内重要宫殿建筑由南及北可分为「前朝」和「内廷」两大部分。

现存前朝区建筑群自午门起始，至乾清门，以太和殿、中和殿、保和殿三大殿为中心，是皇帝举行大典和召见群臣之所。三大殿区域东侧为文华殿，西侧为武英殿建筑群。

Inside the Forbidden City, important buildings from south to north can be divided into two major parts: "Front Court" and "Inner Residence".

Groups of buildings in the Front Court start from the Meridian Gate (Wumen) to Gate of Heavenly Purity (Qianqingmen). Three halls standing at the center are Hall of Supreme Harmony (Taihedian), Hall of Central Harmony (Zhonghedian) and Hall of Preserving Harmony (Baohedian). The three halls were for emperors to hold grand ceremonies or discuss affairs of state with high officials.

午门
Meridian Gate (Wumen)

午门平面呈"凹"字形,是宫城正门 。
午门上部是一座门楼,两翼俗称"雁翅楼",
整座建筑错落有致,左右对应,形似朱雀展翅,故又有"五凤楼"之称。
The Meridian Gate, built on a concave plan resembling the Chinese character "凹",
is the main entrance to the Palace City.

午门重檐飞翘,巍峨壮观(朱雨生/摄)
The spectacular Meridian Gate with doule eaves (Photo by Zhu Yusheng)

从太和门观望午门（袁雪飞／摄）
View of Meridian Gate from Gate of Supreme Harmony (Photo by Yuan Xuefei)

午门——故宫最高处
Meridian Gate – the Highest Point in the Forbidden City

午门位于故宫南北轴线的正南方，也是子午线的午位，故而得名。午门的台基高12米，比10米高的宫墙还高2米，加上门楼，午门总高38米，比太和殿还高。

The Meridian Gate is located in due south on the north-south axis of the Forbidden City, and it is from its location that its name has derived. The abutment of the gate is 12 meters high. Adding the gate tower, the total height reaches 38 meters, even higher than the Hall of Supreme Harmony.

午门怎么走？
How to pass through Meridian Gate

午门中开三门，两旁各有一掖门。一般来说，午门的中门只能皇帝通过，文武百官走东侧门，宗室王公走西侧门。但殿试的前三名即状元、榜眼和探花及第后，可以从午门中门出宫一次，这成为他们一生的荣耀。

The Meridian Gate has five arches, with three of them in the central section and two flanking arches one each side. In general, the arch in the center was exclusively used by the emperor, and all civil and military officials took the east side arch while members of the imperial families took the one on the west side.

内金水河一路蜿蜒流淌，过武英殿转而向东，经太和门、文渊阁，从东南角楼下流出故宫。它不仅为宫内提供了排水通道，还与景山相呼应，形成有山有水、山水协调的审美意象（朱雨生／摄）

The Inner Golden Water River meanders through the Hall of Martial Valor (Wuyingdian), then turns east passing the Gate of Supreme Harmony, the Royal Library, flowing out of the palace through the Corner Tower in the southeast. The Inner Golden Water River is a drainage channel for the imperial palace, while at the same time it resonates with Jingshan Hill to present a mountain-and-river aesthetic appeal (Photo by Zhu Yusheng)

太和门广场
Square of the Gate of Supreme Harmony

太和门前有一片面积约2.6万平方米的广场，内金水河自西向东蜿蜒流过。

河上横架5座石桥，称内金水桥。

广场两侧是排列整齐的廊，一般称为东、西朝房，两侧还有协和门(明代称会极门)和熙和门(明代称归极门)东西对峙。

The square in front of the Gate of Supreme Harmony (Taihemen) encompasses some 26,000 square meters,

through which the Inner Golden Water River winds its way from west to east.

The river is crossed by five stone bridges called the Inner Golden Water Bridges.

On the square are rows of quarters called East Courtyard and West Courtyard.

On the east side of the square lies the Gate of Blending Harmony (Xiehemen),

and on its west side the Gate of Glorious Harmony (Xihemen).

太和门
Gate of Supreme Harmony (Taihemen)

太和门是故宫外朝的正门,也是宫内形制最高的宫门,
为重檐歇山顶,门前列铜狮一对,铜鼎四只。
太和门在明代是"御门听政"的场所,
皇帝在此处接受朝拜与上奏,并处理政事和颁发诏令。

The Gate of Supreme Harmony, the main gate to the Imperial Palace's outer court,
was one of the important places for the emperor to receive ministers, hold court,
handle government issues and issue edicts.

紫禁的含义
The Meaning of "Zi" and "Jin"

　　紫，就是紫微星垣，是世界的中央。中国古代讲究"天人合一"的规划理念，用天上的星辰与都城规划相对应，以突出政权的合法性和皇权的至高性。天帝居住在紫微宫，人间君主自诩为受命于天的"天子"，其居所应象征紫微宫，以与天帝对应。

　　禁，表示皇官是禁地，皇帝办公和日常生活的场所，平民百姓不得擅入。禁，彰显了帝王的特殊权力。

　　The Forbidden City is called "Zi Jin" city in Chinese.

　　"Zi" refers to the Zi Wei star which is the Polaris. According to the capital city planners of ancient China , the layout of the capital city should echo that of the palaces of heaven, so as to highlight the legality of the ruling and supremacy of the royal power. The Heavenly Emperor dwells in the Zi Wei palace in heaven, and as his counterpart, the earthly emperor's palace should symbolize the Zi Wei palace in heaven.

　　"Jin" means to prohibit, showing that the imperial palaces are forbidden areas for the commoners. "Jin" highlights the privilege of the emperor.

太和门前的铜狮子 (站酷/供图)
The copper lion in front of the Gate of Supreme Harmony
(Provided by Zcool.com.cn)

太和殿广场
Square of the Hall of Supreme Harmony

太和殿前的广场面积达3万多平方米，
由太和殿、太和门以及东西两侧的体仁阁、弘义阁等建筑及院墙围合而成。

The square in front of the Hall of Supreme Harmony (Taihedian) covers an area of over 30,000 square meters,
surrounded by the Hall of Supreme Harmony, the Gate of Supreme Harmony,
the Belvedere of Embodying Benevolence (Tirenge) on the east side,
and the Belvedere of Spreading Righteousness (Hongyige) on the west side.

太和殿广场可容纳万人朝拜庆贺（朱雨生／摄）
The square of the Hall of Supreme Harmony has enough space to admit ten thousand people to pay homage to the emperor (Photo by Zhu Yusheng)

太和殿
Hall of Supreme Harmony (Taihedian)

太和殿位于紫禁城中轴线的显要位置，是用来举行国家重大典礼的场所。
它是中国现存最大的木结构大殿之一，
太和殿建筑面阔十一间，进深五间，重檐庑殿顶，屋顶覆盖黄琉璃瓦，檐角走兽多达十个。
在殿前月台上，放置了日晷、嘉量各一个，铜龟、铜鹤各一对，另外还有铜鼎十八座。
殿下是三层汉白玉石雕基座，周围环绕着栏杆，栏杆下设有石雕龙头，用于排水。

The Hall of Supreme Harmony stands in a notable area on the Forbidden City's central axis.
The hall is one of the largest surviving wooden structures in China.
It was where grand state ceremonies were held.

太和殿是紫禁城中最高大的宫殿 (范炳远/摄)
Hall of Supreme Harmony is the largest hall in the Forbidden City (Photo by Fan Bingyuan)

间
Bay

中国古代木构架建筑的基本单元，指两榀相邻梁架之间，由四根柱子围合的面积。在古代中国，建筑间数代表地位等级。根据礼制的规定，十一间、九间的规制唯帝王才配享有。比如，太和殿面阔十一间，冠盖天下建筑。

A bay is a basic unit of timber structured architecture in ancient China. It refers to the area between two adjacent beam mounts enclosed by four columns. In ancient China, the number of bays represented status and ranking.

According to ritual rules, only the emperor is entitled to enjoy buildings with 11 bays or 9 bays. For example, the Hall of Supreme Harmony is 11 bays wide, which ranks the first among all buildings at that time.

铜鹤
Copper Crane

铜龟
Copper Turtle

日晷
Sundial

嘉量
Measurements

太和殿为故宫中最重要的建筑，前朝所在。建筑形制为故宫中轴线建筑中的最高等级。

北京中轴线之上与两侧的建筑规制设计遵循严格等级要求和规定。建筑形制即建筑形态与尺度根据其不同的位置、使用功能、使用者，而在台基、开间进深、结构样式、屋顶形式、材料与色彩、彩画装饰等多方面表现出不同的样式，以体现建筑在整体中的重要性。

The built form of the Hall of Supreme Harmony is of the highest grade among all those on the Imperial Palace central axis. Architectural style and structural design of the buildings on Beijing Central Axis all follow strict ranking requirements and rules.

中和殿的"允执厥中"匾额 (张月军/摄)
The calligraphy tablet hung over the Hall of Central Harmony (Photo by Zhang Yuejun)

三大殿的匾额
Calligraphy tablets hung over the three grand halls

悬挂在太和、中和、保和三大殿的三块匾,内容都从《尚书》中引申而来,分别是"建极绥猷""允执厥中""皇建有极"。其中,中和殿的"允执厥中"意为不偏不倚的中正之道,正是北京中轴线一直遵从的规划和设计准则。

The three tablets hung over the Hall of Supreme Harmony, the Hall of Central Harmony and the Hall of Preserving Harmony bear contents from the *Book of Documents(Shang Shu)*. The Chinese characters "允执厥中" in the tablet in the Hall of Central Harmony means being impartial, firm and upright. This is exactly the planning rules and design criteria closely followed by Beijing Central Axis.

太和殿
Hall of Supreme Harmony

故宫三大殿 (刘雯/摄)
The three major halls of the Forbidden City (Photo by Liu Wen)

大大小小的中轴线
Central axes big and small

除了贯穿故宫中央的中轴主线即所谓中路外，在故宫的东西半区又有着各自的中轴线，分别叫作东路、西路、外东路和外西路。东六宫在东路，东六宫的中轴线是名叫"东二长街"的长巷；西六宫在西路，西六宫的中轴线即"西二长街"；宁寿宫在外东路，慈宁宫在外西路，宁寿、慈宁两宫分别是中轴对称的院落。至于某个具体庭院，比如东西六宫的每一宫，也都是中轴对称的院落。故宫就是由不同级别的院落和中轴线构成的严整的联合体。

Besides the main central axis running through the center of the Forbidden City, mid-areas on the east and west side of the palace also have their own central axes. They are called East Route, West Route, Outer East Route and Outer West Route respectively. When it comes to a specific courtyard, for instance, each of the six palaces on the east and west is arranged symmetrically along their axes. The Forbidden City is actually an orderly compound made up of courtyards and central axes of different ranks.

保和殿
Hall of Preserving Harmony

中和殿
Hall of Central Harmony

三大殿的名称沿革
Changing of the names of the three Halls

明代初建时 During the Ming Dynasty when the three halls were built	明代嘉靖年间起 Since the Ming Emperor Jiajing's reign	清代 During the Qing Dynasty
奉天殿 Hall of Heavenly Worship (Fengtiandian)	皇极殿 Hall of Imperial Zenith (Huangjidian)	太和殿 Hall of Supreme Harmony (Taihedian)
华盖殿 Hall of Magnificence (Huagaidian)	中极殿 Hall of Ultimate (Zhongjidian)	中和殿 Hall of Central Harmony (Zhonghedian)
谨身殿 Hall of Prudence (Jinshendian)	建极殿 Hall of Superb (Jianjidian)	保和殿 Hall of Preserving Harmony (Baohedian)

从故宫中和殿前远眺景山(李少白/摄)
Jingshan Hill at the back as seen from Hall of Central Harmony (Photo by Li Shaobai)

　　故宫内重要宫殿建筑均沿中轴线布局,主要建筑群南北可分为"前朝"和"内廷"两大部分,东西可分为三路。中轴线两侧的武英殿、文华殿、东西六宫等建筑群既呈现出以中轴线左右对称的布局,也构建出自己区域的次级中轴线布局。

　　All the major palaces and buildings within the Imperial Palace are planned symmetrically along the Central Axis. The main architecture groups are divided into two sections: the "Front Court" in the south and the "Inner Residence" in the north. The complexes including the Hall of Martial Valor(Wuyingdian) area, the Hall of Literary Glory (Wenhuadian) area, the Six Eastern Palaces (Dongliugong) area and the Six Western Palaces (Xiliugong) area, as a whole, present symmetrical disposition along the Central Axis. On the other hand, features of secondary-level central axes are also formed within their own areas.

内廷区
Inner Court

中轴线之上的内廷区从乾清门开始，向北到顺贞门为止，这个区域的整体尺度比前朝区更加紧凑，它以乾清宫、交泰殿、坤宁宫后三宫为中心，再向北是御花园。后三宫的东侧分布着奉先殿、斋宫、东六宫等建筑群，西侧分布着养心殿、西六宫等建筑群。后三宫是皇帝与后妃居住、游玩的地方。在后三宫与东西六宫的外围，还分布了宁寿宫、慈宁宫、寿安宫、重华宫、北五所、南三所等建筑群，各有用途。

The Inner Court starts from Gate of Heavenly Purity and ends at Gate of Loyal Obedience (Shunzhenmen) in the north. Its overall scale is even more compact than the Front Court. At the center of the Inner Court is another group of three halls: Palace of Heavenly Purity (Qianqinggong), Hall of Union (Jiaotaidian), Palace of Earthly Tranquility (Kunninggong), and an imperial garden in the north. To the east of the three main halls of the Inner Court are Hall of Ancestral Worship (Fengxiandian), Palace of Abstinence (Zhaigong), and the Six Eastern Palaces. To the west are Hall of Mental Cultivation (Yangxindian), Six Western Palaces. The three halls were where the emperor, empress and imperial concubines lived and spent their leisure time.

乾清门是故宫内廷区的开始（杨春燕/摄）
Gate of Heavenly Purity is where the inner court of the Imperial Palace begins (Photo by Yang Chunyan)

乾清门广场

Square of the Gate of Heavenly Purity

乾清门广场位于保和殿之后，乾清门之前，
联系了前朝和内廷。

The square is located at the rear of the Hall of Preserving Harmony, and in front of the Gate of Heavenly Purity.
It links up the front court and the inner court.

乾清宫

Palace of Heavenly Purity (Qianqinggong)

乾清宫，内廷后三宫之一。

明至清初为皇帝的寝宫。

雍正即位后，乾清宫成为皇帝召见和选派官吏、批阅奏报、处理日常政务和举行重大赐宴活动的场所。

The Palace of Heavenly Purity (Qianqinggong) is one of the three back palaces in the inner court. In the Ming Dynasty until early Qing, it had been the residence of the emperor. When Emperor Yongzheng ascended the throne, the palace then became the emperor's audience hall where he held court, appointed officials, wrote notes and signed documents during council with ministers, handled daily affairs and held grand banquets.

乾清宫曾是皇帝居住和处理政务的地方 (杨春燕／摄)
Palace of Heavenly Purity once served as the emperor's residence and office (Photo by Yang chunyan)

后三宫名称的含义
Names of the back three palaces and their meanings

　　故宫自乾清门向北属于内廷区域，中轴线上依次排列着"后三宫"——乾清宫、交泰殿、坤宁宫。这三座宫殿的名称融合了《易经》和《道德经》的诸多概念。乾、坤、泰都是《易经》中的卦象，乾之象为天，坤之象为地，天地相交为泰。而道家经典《道德经》写道："昔之得一者，天得一以清，地得一以宁……"在古人的宇宙观中，天为父地为母，而皇帝是天子，是天的代表，皇后则为地的代表。于是，皇帝住宿、办公的场所命名为乾清宫，寓意天之高明；皇后起居的场所命名为坤宁宫，寓意地之厚博，乾清宫与坤宁宫之间的宫殿命名为交泰殿，寓意天地交合、康泰美满。

　　Located on the central axis in the inner court are the "back three palaces" : Palace of Heavenly Purity, Hall of Union, and Palace of Earthly Tranquility.

　　According to ancient people's view of the universe, heaven is father and earth is mother; the emperor is the son of heaven, who represents its power, and the empress is the representative of earth. Therefore, the emperor's residence and office was named the Palace of Heavenly Purity, indicating the majesty of heaven. The Empress's residence was named the Palace of Earthly Tranquility, symbolizing the inclusiveness of earth. The hall between them was named the Hall of Union, meaning meeting of heaven and earth, and the blessings of health, prosperity and happiness.

乾清宫内宝座上悬挂"正大光明"匾，皇帝选定并御笔亲书的皇位继承人的名字放置于这块牌匾后面的匣子里 (李少白/摄)
Inside the Palace of Heavenly Purity, the tablet hanging above the throne with a script reading "Zheng Da Guang Ming" (Just and Honorable). The emperor wrote down the name of the successor and put it in a box behind this tablet (Photo by Li Shaobai)

交泰殿
Hall of Union (Jiaotaidian)

交泰殿位于乾清宫和坤宁宫之间，殿名取自《易经》，含"天地交合、康泰美满"之意。
面阔三间，进深三间，四面辟门，屋顶为单檐攒尖顶，
建筑装饰大量用金，如设镏金宝顶，饰龙凤和玺金漆彩画，金扉金锁窗等，显得格外富丽堂皇。

The Hall of Union is located between Palace of Heavenly Purity and Palace of Earthly Tranquility,
meaning the meeting of heaven and earth, and blessings of good health and happiness.

交泰殿，内廷后三宫中轴线上居中建筑，是皇后接受朝贺的场所（杨春燕／摄）
Hall of Union was where the empress received greetings and congratulations (Photo by Yang Chunyan)

交泰殿内景(波音／摄)
Interior of Hall of Union (Photo by Boyin)

宝玺存放处
Place for storing Imperial Seal

　　凡遇元旦、千秋(皇后生日)等重大节日，皇后在交泰殿接受朝贺。乾隆十三年(1748年)，乾隆皇帝把象征皇权的二十五宝玺收存于此，遂为储印场所。现殿内宝座前两侧分别排列着用来储放皇帝宝玺的宝盝。宝座上方高悬康熙帝御笔"无为"匾，殿内东次间设铜壶滴漏，乾隆年后不再使用；西次间设大自鸣钟，宫内时间以此为准。

Emperor Qianlong once put the 25 imperial seals in the Hall of Union and it thus became a repository of imperial seals which were the symbol of imperial power.

坤宁宫

Palace of Earthly Tranquility (Kunninggong)

在明代，坤宁宫是皇后的寝宫，清代改作祭神场所。

清朝皇室每年都要举行大大小小的祭祀，这是皇帝皇后的重要职责之一。

其中一些祭祀需要皇后主持，地点就在坤宁宫中。

坤宁宫建筑面阔九间，进深三间，重檐庑殿顶，顶覆黄色琉璃瓦，饰金凤和玺彩画。

室内东侧两间隔出为暖阁，作为居住的寝室，门的西侧四间设南、北、西三面炕。

In the Ming Dynasty, the Palace of Earthly Tranquility (Kunninggong) was the residence of the empress.

It was converted to a place of worship during the Qing Dynasty.

Every year the imperial family of Qing held numerous sacrificial ceremonies,

which was one of the major duties of the emperor and the empress.

Some of these ceremonies were presided over by the empress in this palace.

坤宁宫，内廷后三宫中轴线上北侧建筑，兼有寝室与祭神场所功能 (杨春燕/摄)
Palace of Earthly Tranquility served as residence as well as a place for offering sacrifice to the gods (Photo by Yang Chunyan)

坤宁宫内景（李少白／摄）
Interior of the Palace of Earthly Tranquility (Photo by Li Shaobai)

祭祀与大婚——坤宁宫的有趣用途
Sacrificial Ceremony and Wedding Ceremony – Interesting uses of Palace of Earthly Tranquility

　　坤宁宫从明代到清初是皇后的居所。在清朝，一些满人的祭祀由皇后主持，因此坤宁宫经过改造成为萨满祭祀等活动的场所。此外，皇帝长大成人结婚时，婚房也设在坤宁宫。不过，由于很多皇帝在即位之前已经大婚，因此有清一代，只有年幼登基的康熙、同治、光绪曾经把坤宁宫作为迎娶皇后的洞房。

　　The palace had been the empress's residence since the Ming Dynasty till early Qing. When the emperor grew to the age of marriage, his wedding room was in this hall. Emperor Kangxi, Emperor Tongzhi and Emperor Guangxu, all ascended to the throne at a young age, and they all used this hall to receive the empress on the wedding night.

御花园
Imperial Garden

御花园位于故宫中轴线的最北部，坤宁宫后方，正南有坤宁门同后三宫相连，
明代称为宫后苑，清代称御花园。
园内青翠的松、柏、竹间点缀着山石，形成四季常青的园林景观。

The Imperial Garden is situated on the northernmost of the Forbidden City's central axis, behind the Palace of Earthly Tranquility.
To due south of the garden is the Gate of Earthly Tranquility to the three back palaces.
Verdant pines, cypresses and bamboos were dotted with artificial hills and rocks,
forming an evergreen landscape.

御花园百年连理枝（杨春燕／摄）
Imperial Garden's century-old trees with branches intertwined (Photo by Yang Chunyan)

天一门是由青砖砌成的单洞券门，是皇城内为数不多的青砖建筑之一。天一门的取名来自古代文化中的"天一生水"概念，水在中国文化中有万物之源的意义，同时也反映了避火防灾的愿望。

Tianyi Gate, built of grey bricks, is an arch gate with one gateway, one of the few grey-brick structures in the Imperial City. The name of the gate comes from the ancient Taoist conception of "Tian Yi Sheng Shui", or heavenly blessed water generates everything, as in the context of Chinese culture, water is the source of all things. It also reflects the desire to avoid fire.

御花园天一门 (阮旭红／摄)
Tianyi Gate to the Imperial Garden (Photo by Ruan Xuhong)

故宫白玉兰绽放 (朱雨生／摄)
Magnolia Bloom (Photo by Zhu Yusheng)

静憩斋杏花开 (刘雯／摄)
Apricot Bloom (Photo by Liu Wen)

故宫里的花园
Gardens in the Forbidden City

故宫内有四座大小不等的花园，分别是御花园、慈宁宫花园、建福宫花园、宁寿宫花园(即乾隆花园)。其中以坐落于中轴线上的御花园面积最大，也最为重要。

There are four gardens of various sizes within the Forbidden City. They are respectively the Imperial Garden, Garden of Compassion and Tranquility, Garden of Established Happiness, Garden of Tranquil Longevity (also known as Qianlong Garden). The largest and the most important is the Imperial Garden sitting on the central axis.

名称 Name	始建时间 First Built Time	花园及其宫殿的功能 Use of the garden and its palace
御花园 Imperial Garden	1420年(明永乐时期) 1420 (Yongle Era, Ming Dynasty)	御花园体现了古代宫廷"前宫后苑"的格局，是供帝王后妃游玩休憩的场所。 Imperial Garden showcases the layout of "Palace in front, Garden at the back". The garden is a private retreat for the emperors, empresses and consorts.
慈宁宫花园 Garden of Compassion and Tranquility	1538年(明嘉靖时期) 1538 (Jiajing Era, Ming Dynasty)	慈宁宫是前代皇后、皇贵妃的居所，清代孝庄文皇后、孝圣宪皇后都曾在此居住和礼佛。 This garden was the residence of empresses and concubines of the previous emperors. Empress Xiaozhuangwen and Empress Xiaoshengxian of the Qing Dynasty had lived here and practiced Buddhism.
建福宫花园 Garden of Established Happiness	1742年(清乾隆时期) 1742 (Qianlong Era, Qing Dynasty)	建福宫为乾隆所建，初建时本意"备慈寿万年之后居此守制"，后成为珍奇文物的收藏宝库。 Emperor Qianlong built the palace with the intention to "observe the ritual of filial piety and stay here to mourn his mother when she would pass away after living a long life". Later the palace became a repository for treasures and antiques.
宁寿宫花园 Garden of Tranquil Longevity	1776年(清乾隆时期) 1776 (Qianlong Era, Qing Dynasty)	宁寿宫是乾隆为自己退位后生活而打造的宫苑区。 Garden of Tranquil Longevity is a complex built by Emperor Qianlong in anticipation of his retirement.

钦安殿

Hall of Imperial Peace (Qin'andian)

钦安殿立于故宫中轴线之上、御花园的正中，是御花园的核心建筑，它的用途是供奉玄天上帝。

钦安殿建筑面阔五间，进深三间，重檐盝顶，黄琉璃瓦顶。

整个建筑坐落在汉白玉石单层须弥座上，殿前出月台，四周包围着穿花龙纹汉白玉石栏杆。

The Hall of Imperial Peace sits on the Forbidden City's Central Axis
at the center of the Imperial Garden.
It served as a place enshrining God of the North.

位于御花园正中的钦安殿（站酷／供图）
Hall of Imperial Peace in the Imperial Garden (Provided by Zcool.com.cn)

　　钦安殿是紫禁城内保留下来的为数不多的明代中期建筑之一，殿顶造型独特、精工细致，配上金黄色的琉璃筒瓦、华丽的和玺彩画和生动形象的云龙雕刻，富丽堂皇、壮观庄严。

　　The Hall of Imperial Peace is one of the rare historic buildings in the Forbidden City preserved from the mid Ming Dynasty.

千秋亭位于御花园西侧，与万春亭沿中轴线对称分布 (范炳远／摄)
Qianqiu Pavilion on the west side of the Imperial Garden (Photo by Fan Bingyuan)

万春亭位于御花园东侧，与千秋亭沿中轴线对称分布 (牛飞／摄)
Wanchun Pavilion on the east side of the Imperial Garden (Photo by Niu Fei)

千秋亭藻井彩画 (范炳远／摄)
Colored painting of the caisson ceiling of Qianqiu Pavilion (Photo by Fan Bingyuan)

雪中神武门　城基为青灰色歇山顶，下层单檐单昂五踩斗拱，上层单拱重昂七踩斗拱，未杨间饰暗线式镂金菱手彩画，正楼楼正庭檐金额字满汉文"神武门"区须。顶度黄色琉璃瓦（李建辉/摄）
Gate of Divine Prowess in the snow (Photo by Li Jianhui)

神武门
Gate of Divine Prowess (Shenwumen)

神武门是故宫正北门，与南面的端门一起，限定了故宫中轴线的南北端点。
在明清两代，皇后行亲蚕礼时会出入神武门。神武门下建有城台，辟三券门，上建城楼。
城楼建在汉白玉基座上，面阔五间，进深一间，四周围廊，以汉白玉石栏杆环绕。

The Gate of Divine Prowess is the northern entrance to the Forbidden City.
This gate in the north and the Upright Gate in the south are the north and south end of the central axis of the Forbidden City.
During the Ming and Qing dynasties, the empress walked through this gate and led the noble ladies in the palace to make sacrifice to the silkworm goddess. The gate abutment has three arched gateways with a tower built on it.

中国最大的古代文化艺术博物馆
Largest museum in China for ancient culture and art

　　故宫南北长961米，东西宽753米，四面环有高10米的城墙，城墙外建有宽52米的护城河，城墙四周各设城门一座，四角各设角楼。1925年，故宫博物院于此成立，自此明清两代皇宫禁地成为对公众开放的博物馆，实现其功能的公众化转变。

　　The Forbidden City is 961 meters long from north to south and 753 meters wide from east to west, surrounded by 10-meter-high city walls. Outside the city walls is a 52-meter-wide moat. There is a gate in each of the four sides of the city walls, and on each of the four corners sits a Corner Tower. In 1925, the Palace Museum was established and opened to the public.

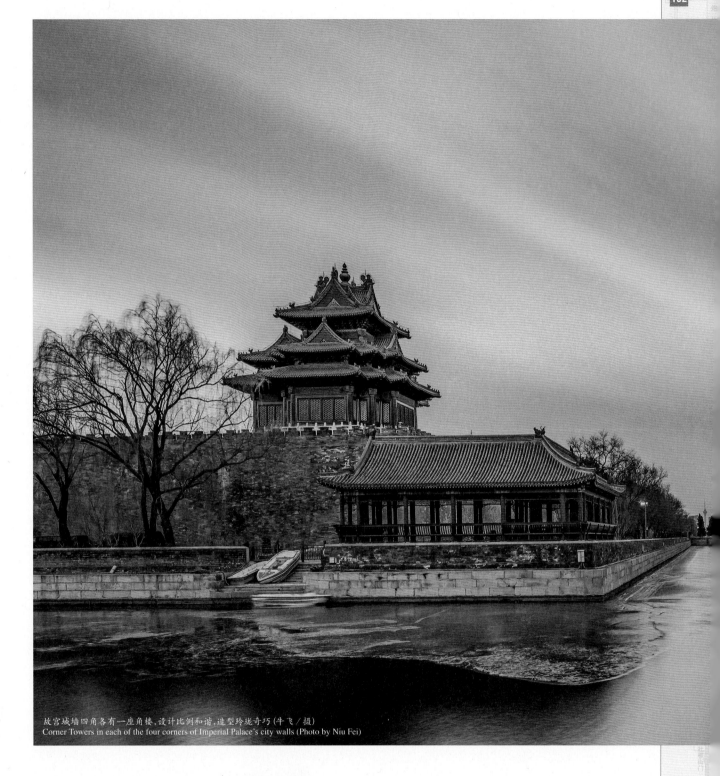

故宫城墙四角各有一座角楼，设计比例和谐，造型玲珑奇巧（牛飞／摄）
Corner Towers in each of the four corners of Imperial Palace's city walls (Photo by Niu Fei)

角楼
Corner Towers

故宫四角分别设置了角楼，
它与城墙、城门楼及护城河同属于皇宫的防卫设施。

A Corner Tower was built in each of the four corners of the Forbidden City.
Corner towers are part of the defence facilities along with the city walls, gate towers and the moat.

角楼不是真正的楼
The Corner Tower is not a multi storied building

　　角楼俗称有"九梁十八柱、七十二条脊",是一座四面"凸"字形平面组合的多角建筑,从外面观看,它有三重檐飞翘的屋檐,却只有一层屋身,似楼非楼。

　　The Corner Tower is commonly known as having "9 girders, 18 columns and 72 ridges". It is a polygonal building with a combination of four convex-shaped planes. Viewed from outside, it has triple eaves, but only one floor.

太庙
Imperial Ancestral Temple

民国时期的北京太庙 (fotoe.com／供图)
Imperial Ancestral Temple during the Republican period (Provided by fotoe.com)

　　太庙坐北朝南,位于明清皇城之内,在宫城紫禁城的东南角、天安门东侧,是明清两代皇家的祖庙。太庙的位置遵从《周礼·考工记》里"左祖右社"的布局。太庙的平面结构为南北略长、东西略窄的长方形, 总面积约16.9万平方米。

　　Sitting in the north and facing south, the Imperial Ancestral Temple is situated at the southeastern corner inside the Ming-Qing imperial city, to the east of Tian'anmen Gate. It is the ancestral temple of the imperial families of the Ming and Qing dynasties.

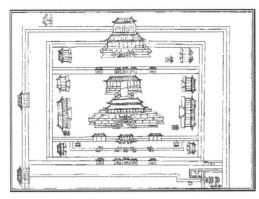

太庙总图（《钦定四库全书》）
Plan of Imperial Ancestral Temple
(*Imperial-Endorsed Complete Library in Four Sections*)

大戟门将太庙的中区和南区分开，两侧为小戟门 (杨春燕／摄)
The Halberd Gate (Dajimen) divides the central zone and southern area of the
Temple, with the two small Halberd Gates on each side (Photo by Yang Chunyan)

　　太庙建筑群由内垣墙和外垣墙围合，形成两层环套式院落区域，它的布局是中国宗庙祭祀传统中的"合祀"制度的反映。以内垣墙为边界，向内是太庙的核心院落，主要的祭祀建筑都位于院落内，建筑群采用了中轴对称的布局方式。核心院落以外到外垣墙之间是太庙的外院范围，种植了数量庞大的古树，营造出祭祀先祖的幽静氛围。

　　The temple complex is surrounded by inner walls and outer walls. Its layout is a reflection of "combined offering" in Chinese tradition of ancestral veneration. The inner walls are the boundary of the temple's principle courtyard where main sacrificial buildings are located.

大戟门外的金水河与金水桥 (金东俊／摄)
Golden Water River and Golden Water Bridge outside the Halberd Gate (Dajimen) (Photo by Jin Dongjun)

　　太庙的主体建筑为享殿、寝殿和祧庙。享殿正南是大戟门。大戟门外金水河自西向东流淌，河上架金水桥，桥北面东西各有一座六角井亭，桥南面为神厨与神库。再往南是五彩琉璃门，门外的东南有宰牲亭、治牲房和井亭等。

　　Main buildings of the Imperial Ancestral Temple are the three main halls. In due south of the Hall of Worship of Ancestors stands the Halberd Gate. The Golden Water River outside the Halberd Gate flows from west to the east. The river is crossed by the Golden Water Bridge.

享殿
Hall of Worship of Ancestors (Front Hall)

太庙前殿即享殿，是明清两代皇室举行祭祖大典的场所。其建筑规制在现存明清建筑中等级最高。
享殿在明初面阔九间，而后改为面阔十一间，重檐庑殿顶，是皇帝祭祀先祖时行礼的地方。
从建筑构造、彩画装饰、用料等方面，享殿都充分地体现出其重要的建筑地位和价值。

The Front Hall of the Temple is the Hall of Worship of Ancestors
where large-scale ceremonies were held in honor of the imperial family's ancestors during the Ming and Qing dynasties.
Its built form is of the highest ranking among the existing Ming and Qing buildings,
almost equal to that of Hall of Supreme Harmony.
The hall served as the place for the emperors to pay homage to their ancestors.

太庙享殿（杨春燕／摄）
The Hall of Worship of Ancestors of the Imperial Ancestral Temple (Photo by Yang Chunyan)

太庙享殿内部(杨春燕/摄)
Interior of the Hall of Worship of Ancestors (Photo by Yang Chunyan)

祭祖,祭的都是谁?
Sacrifices were offered, but to whom?

在中国古代的宗法制度中，对于祖先的崇拜祭祀分为三类祖先神：近祖、远祖和始祖，也分别被称为祢祖、祧祖和太祖。中国古代的氏族通常是由同一个男子五代以内(含第五代)的男性子孙及其配偶和未出嫁的女性子孙组成。因而，往上四代(即高祖父、曾祖父、祖父、父)为近祖，第五代以外的祖先为远祖，最初的远祖即为始祖。

In ancient China's lineage system, ancestors are classified into three types: forefathers, remote ancestors and earliest ancestors. In ancient China, a clan is formed by five generations (including the fifth generation) of one man's male descendants, their spouses, and unmarried female descendants. Therefore, the previous four generations are forefathers, and distant ancestors refer to ancestors beyond the fifth generation. The most distant generation is the earliest ancestor.

"迁主毁庙" 制度
"Changing temple owner" System

太庙中帝王祖先的排位十分讲究。由于氏族内的
祖先只包含四代，因而对应到宗庙建筑上，就出现了四
亲庙，分别是显考庙（高祖庙）、皇考庙（曾祖庙）、王考
庙（祖庙）、考庙（父庙）。每当皇帝去世，原来的氏族祖
先位置就要向上调整庙位，调整的结果，原高祖的显考
庙就不存了，这就是所谓的"毁庙"，即神主改迁他处，
换新的牌位。

The Imperial Ancestral Temple followed rules of great
particularity in arranging the ancestral tablets. Since the
ancestors within the clan only include four generations, there
appeared corresponding ancestral halls for four generations
respectively for the forefather, the great grandfather, the
grandfather and the father. When an emperor passed away,
positions of those ancestors had to be adjusted. As a result, the
forefather's tablet had to be removed. Therefore, the so-called
"changing temple owner" means moving the ancestral tablets
to other places to make room for the newcomer.

寝殿
Resting Hall (Middle Hall)

太庙中殿即寝殿,是平时供奉历代皇帝、皇后牌位的地方,殿内祖宗牌位同堂异室安置。

到清末时,寝殿里供奉了十一代皇帝和皇后的神主牌位。

在岁末享殿举行祭祀的前一天,人们将牌位移至享殿,安放于神座之上,祭祀完毕后,再奉回中殿。

中殿为九间四进,黄琉璃筒瓦、单檐庑殿顶。

The Middle Hall, or the Resting Hall, is where memorial tablets of deceased emperors and empresses were enshrined.
In the late Qing Dynasty,
it housed ancestral tablets of eleven generations of emperors and empresses.

太庙寝殿(站酷／供图)
The Resting Hall of the Imperial Ancestral Temple (Provided by Zcool.com.cn)

桃庙
Tiao Miao (Back Hall)

太庙后殿即桃庙，
是供奉皇帝远祖牌位的地方，殿内陈设与中殿基本相同。
后殿的形制也与中殿基本一致，都是九间四进。

The Back Hall, or Tiao Miao,
was where memorial tablets of the remote ancestors of the emperors were enshrined.
Furnishings and the built form of the Back Hall are basically consistent to those of the Middle Hall.

太庙桃庙（金东俊／摄）
The Tiao Miao of the Imperial Ancestral Temple (Photo by Jin Dongjun)

太庙作为皇帝祭祀祖先的宗庙,其传统可上溯到原始氏族制度中通过祈求祖先神灵以保佑氏族平安的习俗。这种祖先崇拜的方式逐渐以礼制形式被赋予了法的性质,将祭祀权和政治权联系在了一起,逐渐形成等级森严的宗法制度。中国历朝历代都将祖先祭祀视为重要的大事,为其修建祭祀建筑。

中华人民共和国成立后,太庙改作北京市劳动人民文化宫,成为公众文化活动场所。

Imperial Ancestral Temple was where the emperors worshiped their ancestors. The tradition can be traced back to primitive clan system, when people prayed for ancestors' blessing to protect the clans. As it developed, this kind of ancestral worship was given a lawful nature in the form of rites linking up worship rights and political rights, and evolved into a rigid hierarchical lineage system. In all Chinese dynasties throughout history, imperial ancestor worshipping had been such a significant event that sacrificial sites were constructed to serve the purpose.

After the founding of the People's Republic of China, the Imperial Ancestral Temple was transformed into a public space for cultural and leisure activities, under the name of Beijing Working People's Cultural Palace.

昭穆制度
"Zhao-Mu" Ranking System

明清两代,太庙陈列排位时遵循严格的昭穆制度,按照"父为昭,子为穆"的原则分成左右不同两列,左为昭,右为穆。父与子不同列,而祖孙却异辈同列。

During the Ming and Qing dynasties, the memorial tablets of the ancestors in the imperial temples were arranged according to the strict Zhao-Mu system. According to the principle of "the father is Zhao, the son is Mu", they were placed in two different columns with Zhao on the left and Mu on the right. The father and the son should not be in the same column, while the grandfather and the grandson were placed in the same column.

社稷坛
Altar of Land and Grain

《唐土名胜图会》中的社稷坛 (fotoe.com／供图)
Altar of Land and Grain as portrayed in the *Pictorial Book of China's Scenery* (Provided by fotoe.com)

社稷坛建于明永乐十八年（1420年），整体布局为南北稍长的不规则长方形，有内外两重坛垣。在中国传统文化中，社稷坛是微缩的天下，体现了古人的世界观和宇宙观，反映了古人对土地的依赖和尊崇，是中国社会对农业乃立国之根本观念的体现。1914年开始，社稷坛开始作为城市公园对公众开放，成为北京老城内第一处转变为公园的皇家坛庙建筑，开启了北京中轴线近现代公众化的转变历程。

Altar of Land and Grain was built in the 18th year (1420) of Yongle era of the Ming Dynasty. The entire layout is an irregular rectangular, slightly longer from south to north. In Chinese traditions, Altar of Land and Grain is a miniature of the universe, a reflection of ancient people's view of the world, and their reliance on and respect for the land. As of 1914, Altar of Land and Grain has been open to the public as a city park, the first imperial altar site in the old city of Beijing that was transformed into a park.

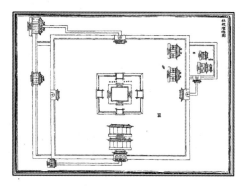

社稷坛总图(《钦定四库全书》)
Plan of Altar of Land and Grain
(*Imperial-Endorsed Complete Library in Four Sections*)

社稷坛位于明清皇城之内、天安门西侧,与太庙形成东西对应的关系,二者的位置体现了北京中轴线"两翼对称"的特色。社稷坛是明清两代皇帝在每年春秋仲月上戊日举行祭祀社神、稷神仪式,祈求五谷丰登、国泰民安的地方。

Altar of Land and Grain was where the emperor held the sacrificial ceremonies in honor of god of land and god of grain on the second month of spring season and autumn season every year.

社稷坛今为中山公园(顾彩华／摄)
Altar of Land and Grain – Today's Zhongshan Park (Photo by Gu Caihua)

中山公园古柏(顾彩华／摄)
An old Cypress inside Zhongshan Park (Photo by Gu Caihua)

祭坛
Altar for Worship

社稷坛祭坛俗称五色土坛。现存五色土坛为三层，四面居中各设置了台阶，用青白石砌筑。
坛台之上按照方位铺设五色土：东青、南红、西白、北黑、中黄，中央立"社主石"，亦称"江山石"。
以这种特殊的方式象征国家辽阔疆域，表达江山永固的寓意。

The Altar of Land and Grain is commonly known as the altar of five-colored soils.
The soils are arranged on the altar in five directions: green in the east,
red in the south, white in the west, black in the north, and yellow in the middle.
At the center stands a stone representing the country.

五色土坛与拜殿 (视觉中国／供图)
The altar of five-colored soils and the Worship Hall (Provided by Visual China)

来自五方的土
Soils from five directions

明初修建社稷坛后，祭坛所用的五色土由各地运来，居中的黄土采自河南，居东的青土采自山东，居南的赤土从广东、广西运来，居西的白土来自陕西，居北的黑土则选自北京。

When the Altar of Land and Grain was completed in early Ming, the five colors of soils for the worshipping altar were shipped from different parts of the empire, with yellow in the middle from Henan, green in the east from Shandong, red in the south from Guangdong and Guangxi, white in the west from Shaanxi and black in the north from Beijing.

"社稷"是"太社"和"太稷"的合称，社是土地神，稷是五谷神，两者是农业社会最重要的根基。在古代中国，不仅首都有国家的祭坛，地方各级城市也都有祭祀社稷的场所。

由于土地与五谷是中国作为传统农业社会的立国之本，因此对它们的崇拜极为重视。祭祀社、稷的活动逐渐演变为祈祷国家根基稳固、繁荣昌盛、疆域完整的意义。

The Altar of Land and Grain is called "Sheji Tan" in Chinese. "Sheji" is a merged term of "She", the God of Land, and "Ji", the God of Grain, which are principal foundation of an agricultural society. A state altar was set up in the capital city, meanwhile other cities of various sizes had their own places dedicated to the worship of the gods of land and grain.

唐花坞(金东俊／摄)
Tanghuawu (The greenhouse) (Photo by Jin Dongjun)

保卫和平石牌坊(陆岩／摄)
Stone Arch of Safeguarding Peace (Photo by Lu Yan)

社稷坛与太庙左右相对，是明清两代北京老城内极为重要的国家礼制建筑。明清两代，社稷坛一直是皇家用以供奉、祭祀土地神和五谷神的场所。虽然历朝历代的社稷制度有所不同，但其主要功能从未改变。为表达"非土不立，非谷不食，王者以土为重，为天下求福报功"的意愿，每年春秋仲月上戊日清晨，皇帝都会在社稷坛举行大祭，如遇出征、班师、献俘等重要事件，也在此举行社稷大典。

The Altar of Land and Grain on the left and the Imperial Ancestral Temple on the right are of symmetric disposition. They are important ritual architectures inside the Imperial City of Beijing.

拜殿今为中山堂(金东俊／摄)
Worship Hall – Today's Zhongshan Hall (Photo by Jin Dongjun)

拜殿
Worship Hall

拜殿位于社稷祭坛正北，
面阔五间，进深三间，周围不设廊。
祭祀日遇风雨时拜殿是行礼的场所，无风雨时则用于设置御幄。

The Worship Hall is situated in the due north of the Altar of Land and Grain.
On the day of the sacrificial ceremony, if there was wind and rain, the Worship Hall would be used as the place for salutation;
if there was no wind or rain, it would be used to set up the imperial tent.

戟门
Jimen Gate

戟门位于拜殿北侧，
是社稷坛中轴线上的核心建筑。
建筑面阔五间，进深两间，明次间为大门。

The Jimen Gate,
located to the north of the Worship Hall,
is a key building on the central axis of the Altar of Land and Grain.

社稷坛戟门（牛 飞／摄）
Jimen Gate of the Altar of Land and Grain (Photo by Niu Fei)

格言亭(金东俊／摄)
Motto Pavilion (Photo by Jin Dongjun)

北京市民公园的先河
First civic park in Beijing

民国时期，社稷坛改作中山公园对外开放，成为北京第一座由皇家禁地开放而成的城市公园，开辟了北京市民公园的先例。在皇家坛园基础上营建城市公园，创造性地解决了保护和利用的难题——既保护延续明清时期皇家祭坛、收集保护散落的历史遗存，又开辟满足市民需求的各类设施和空间，是近代中国文物保护实践的重要尝试。

During the Republican period, the Altar of Land and Grain was open to the public as Zhongshan Park. It was Beijing's first civic park established by opening the once forbidden imperial area, a pioneering example of its kind.

天安门广场是今天国家级重大庆典和活动的举办场所(闫立军／摄)
Tian'anmen Square is where major national celebrations and memorial activities take place (Photo by Yan Lijun)

汲古开新

Drawing from the Old to Inspire the New

天安门广场建筑群
Tian'anmen Square Complex

天安门前的汉白玉华表(顾彩华／摄)
Ornamental Pillar of white marble in front of
Tian'anmen Gate (Photo by Gu Caihua)

天安门广场建筑群位于北京中轴线中心段落,北邻故宫,南望正阳门。这一段落分布着天安门城楼、天安门广场、人民英雄纪念碑、毛主席纪念堂、中国国家博物馆以及人民大会堂。这些建筑虽然大多是中华人民共和国成立后的建设成就,但从规划布局、建筑体量与立面设计上看,延续了传统北京中轴线居中对称的原则,充分体现出建筑规划设计者对北京中轴线建筑景观秩序的遵循与增强。

天安门城楼位于这个段落的北端,其南侧紧邻外金水桥、外金水河历史水系、石狮子、华表等,它们均以北京中轴线左右对称设置。其中,天安门城楼是明清两代皇城的正门,也是中华人民共和国的象征性建筑之一,站在天安门城楼上,向南可以俯瞰天安门广场全貌。

在天安门广场内,人民英雄纪念碑、毛主席纪念堂一北一南分布,两座建筑均坐南朝北;中国国家博物馆和人民大会堂则以中轴线左右对称,分别位于广场东侧和西侧,面对广场。中国国家博物馆和人民大会堂的建筑面积相差较大,建筑设计师为了加强以中轴线对称的视觉效果,在设计时特地尽量保持两栋建筑体量、高度与立面形态的对称性。

从明代开始,现在的天安门广场及其周边区域就是国家机构所在地,中华人民共和国成立后,这里更成为政治活动的中心。广场的形态格局、广场上重要历史纪念物的选址、相关历史建筑的建设,都尽量考虑以中轴线严格对称,体现出传统的北京中轴线思想对现代设计的影响。作为一组以北京中轴线建构的有机整体,天安门广场建筑群蕴含了强烈的纪念性和礼仪性,它们既是北京中轴线近现代公众化转变最具代表性的实例,又延续了长期以来作为城市礼仪核心的文化传统。

The Tian'anmen Square Complex lies at the middle section of Beijing Central Axis, connecting the Forbidden City in the north and the Zhengyangmen Gate in the south. This section of the Central Axis contains the Tian'anmen Gate Tower, the Tian'anmen Square, the Monument to the People's Heroes, the Chairman Mao Memorial Hall, the National Museum of China, and the Great Hall of the People. These structures embody construction accomplishments of 1950's-1970's, while following the traditional principle of respecting central location and symmetrical layout, observing and reinforcing the order of architectural landscape developed along Beijing Central Axis.

The Tian'anmen Gate Tower stands at the north edge of the section. To its south are the Outer Golden Water Bridges (Wai Jinshui Qiao), the historical water system of the Outer Golden Water River (Wai Jinshui He), the Stone Lions and the Ornamental Pillars (Huabiao), which are arranged symmetrically on both sides of Beijing Central Axis. Among them, the Tian'anmen Gate Tower is the main gate to the Imperial City of the Ming and Qing dynasties and an iconic monument to symbolize the founding of the People's Republic of China. On the gate tower, one can overlook the whole of the Tian'anmen Square in the south.

The Monument to the People's Heroes and the Chairman Mao Memorial Hall are arranged from north to south within the Tian'anmen Square, the National Museum of China and the Great Hall of the People lie on the east and west sides of the square respectively in symmetry along the Central Axis, facing each other across the square. Due to the size difference between the two structures, symmetrical designs were applied to their volumes, heights and facades to the extent possible, in order to maintain and enhance the visual symmetry of the Central Axis.

The present-day Tian'anmen Square and its surrounding area has long been the seat of national government offices since the Ming Dynasty, and they have further become the center of political activities and ceremonies today. The form and layout of the square as well as locations of important monuments and related buildings within the square give full consideration to the strict symmetry of the Central Axis, manifesting the influence of Beijing Central Axis planning conception over modern designs. As an integrated ensemble on Beijing Central Axis, the Tian'anmen Square Complex displays a prominent monumental and ceremonial nature. The Complex is the most representative example to witness the transition of Beijing Central Axis toward public access in modern times, while carrying on its cultural tradition as the longstanding core of the city's etiquette.

天安门广场建筑群
Tian'anmen Square Complex

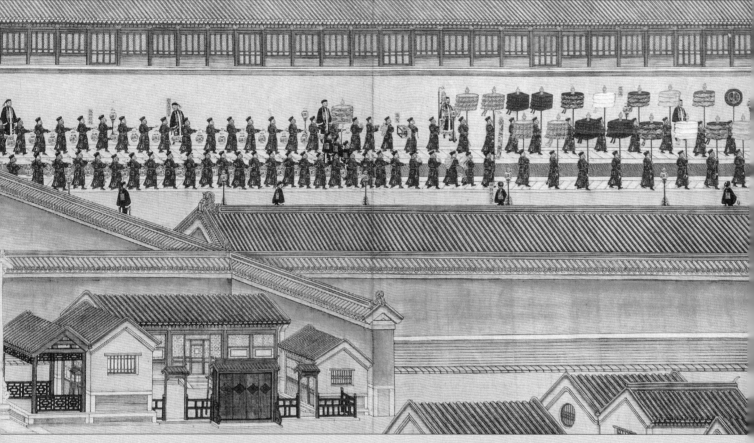

千步廊与御道是天安门广场的前身 (fotoe.com／供图)
Thousand-Step Corridor with the Imperial Passage in the middle was the predecessor of the Tian'anmen Square (Provided by fotoe.com)

　　天安门广场建筑群坐落在北京中轴线上，南至正阳门城楼，北邻天安门城楼，东有中国国家博物馆，西拥人民大会堂，在广场的中轴线上，中部为人民英雄纪念碑，南部是毛主席纪念堂。

　　The Tian'anmen Square, located on Beijing Central Axis, is encompassed by the Zhengyangmen Gate Tower in the south, the Tian'anmen Gate Tower in the north, the National Museum of China in the east, and the Great Hall of the People in the west. Standing on the central axis of the square are the Monument to the People's Heroes in the middle, and the Chairman Mao Memorial Hall in the south.

天安门广场 (陆岗 ／ 摄)
Tian'anmen Square (Photo by Lu Gang)

天安门广场建筑群 (阮旭红 ／ 摄)
Tian'anmen Square Complex (Photo by Ruan Xuhong)

天安门城楼
Tian'anmen Gate Tower

作为明清两代皇城的正门，
天安门城楼始建于明永乐十五年(1417年)。

As the main entrance to the Imperial City of the Ming and Qing dynasties,
the Tian'anmen Gate Tower was first constructed in 1417 (the 15th year of the Yongle era, Ming Dynasty).

天安门城楼两侧观礼台（金东俊 摄）
The viewing stands on both sides of the Tian'anmen Gate Tower (Photo by Jin Dongjun)

城楼上宣告清朝灭亡
The Fall of the Qing Dynasty Announced at the Tian'anmen Gate Tower

　　1911年辛亥革命爆发,1912年2月12日,清朝隆裕太后以宣统皇帝的名义发布退位诏书,正式宣告清朝的灭亡,标志着中国长达2000多年封建王朝统治的终结。

　　Following the 1911 Revolution, on February 12, 1912, Empress Dowager Longyu announced from the Tian'anmen Gate Tower abdication of Emperor Xuantong which marked the fall of the Qing Dynasty. This event marks the termination of feudal rule that had sustained in China for more than two thousand years.

天安门是中华人民共和国的象征，每天都会在此举行升旗与降旗仪式 (陆岗／摄)
Tian'anmen Gate is the symbol of the People's Republic of China. The national flag raising and lowering ceremonies are held daily here (Photo by Lu Gang)

近现代时期，天安门作为一系列影响中国历史发展走向的重大事件的发生地，见证了中国长达2000多年封建王朝时代的结束，也见证了中华人民共和国诞生的光辉一页。由此天安门成为中华人民共和国的国家象征，其形象凝固于中华人民共和国国徽之上，被赋予了新的时代意义。

In modern times, the Tian'anmen Gate was where a series of major events occurred that have changed the trajectory of Chinese history. It witnessed the termination of the imperial rule of dynasties that lasted for more than 2,000 years and the glorious founding of the People's Republic of China. Tian'anmen Gate is hailed as a national symbol of China with its image sealed in the National Emblem.

外金水桥
Outer Golden Water Bridge

外金水桥北倚天安门城楼，南邻天安门广场，由五座三孔拱券式汉白玉石桥构成。
五座石桥分别与天安门城楼的五个门洞相对。
外金水桥的栏杆雕琢精美，形似条条玉带，与古朴的华表和雄伟的石狮构成天安门前巍峨壮丽的景色。
由于故宫内也有金水桥，因而以内、外金水桥在名称上做区分。

With the Tian'anmen Gate Tower in the north and the Tian'anmen Square in the south,
the Outer Golden Water Bridge is a set of five white marble bridges, each of a trip-arched structure.
These five stone bridges correspond to the five archways of the Tian'anmen Gate Tower respectively.
The railings of the bridges are beautifully carved and shaped like strips of jade belts.

五座外金水桥与天安门城楼的五个门洞相对（李少白／摄）
The five stone bridges correspond to the five gateways of the Tian'anmen Gate (Photo by Li Shaobai)

中华人民共和国万岁

不能随便走的桥
The Bridge of Privileged Access

外金水桥正中间的桥是蟠龙雕花柱,桥面最宽,在皇权时代此桥称为"御路桥",只限天子行走;"御路桥"两旁的桥叫"王公桥",只许宗室亲王行走;"王公桥"再左右的桥叫"品级桥",准许三品以上的文武大臣行走。而太庙和社稷坛门前的桥称为"公生桥",准许四品以下的官员行走。

Among the five arch bridge structures of the Outer Golden Water Bridge, the one in the very middle is the widest, with balustrades decorated with carved dragon and cloud patterns. This bridge, once called the "Imperial Road Bridge" during the times of imperial rule, was exclusively reserved for the emperor. The two bridges on either of its nearest sides, known as the "Bridges for Nobilities", allowed access by princes only. And the two on its farthest sides, named as the "Bridges for High-rank Officials", allowed access by officials ranked third and above. The bridges in front of the Imperial Ancestral Temple and the Altar of Land and Grain, both called "Bridge of Offcials Justice", were for officials ranked fourth and below.

天安门广场

Tian'anmen Square

天安门广场南北长880米、东西宽500米、总面积达44万平方米，
地面采用浅粉色花岗岩铺装。
广场地面铺装与绿化布局均以北京中轴线呈对称布置。
今日的天安门广场是国家大型庆典活动的举办地。

The Tian'anmen Square is 880 meters long from north to south and 500 meters wide from east to west,
totaling an area of 440,000 square meters.
It has a floor paved with light pink granite slates.
Both its pavement and greening layout are arranged symmetrically along Beijing Central Axis.
Today, the Tian'anmen Square is where major national ceremonies and celebrations are held.

天安门广场建筑群依照北京中轴线规划布局 (视觉中国／供图)
The Tian'anmen Square Complex has been built in line with the overall plan of Beijing Central Axis (Provided by Visual China)

天安门广场的形成
Forming of the Tian'anmen Square

天安门广场形成于20世纪开始的对该地区原有"T"字形宫廷广场的改造。1913年，北洋政府开辟了天安门前东西大道；其后，又拆除了天安门前东西三座门两侧的宫墙，初步形成一个小广场。1958—1959年，为迎接中华人民共和国成立十周年大庆，结合十大建筑的建设首次对天安门广场进行大规模扩建。第二次大型改造则伴随着毛主席纪念堂的建设，于1977年形成今日宏大而开敞的规模。

The Tian'anmen Square was created out of the renovation of the original T-shaped court square that began from the early 20th century. In 1913, the Beiyang (northern warlords) government opened up the east and west avenues in front of the Tian'anmen Gate. It subsequently removed the palace walls in front of the Tian'anmen Gate, giving rise to a small square initially. In 1958 and 1959, when the "ten great buildings" project was launched in preparation for the grand celebration of the 10th anniversary of the People's Republic of China, the Tian'anmen Square underwent the first ever massive expansion. The second massive renovation took place along with the construction of the Chairman Mao Memorial Hall in 1977, which gave birth to its magnificent and spacious scale today.

人民英雄纪念碑
Monument to the People's Heroes

人民英雄纪念碑矗立于北京中轴线上，主面朝北，背面朝南。

碑基占地面积3000多平方米，碑高37.94米，由1.7万块坚固美观的花岗石和汉白玉砌成。

纪念碑碑身正面最醒目的部位，装着一块巨大的花岗石，上面镶刻着镏金的"人民英雄永垂不朽"8个大字。

另外，在人民英雄纪念碑下层大须弥座的束腰部分，还镶嵌有10块汉白玉浮雕，浮雕内容反映了重大历史事件。

The Monument to the People's Heroes stands on Beijing Central Axis, facing north.

The pedestal occupies an area of more than 3,000 square meters, and the monument is 37.94 meters high,

made from 17,000 pieces of solid and beautiful granite and white marble.

The most prominent part of the front side is a huge granite piece in which eight gilded Chinese characters are inscribed,

reading "The people's heroes are immortal".

The waist of the great sumeru pedestal at the lower tier of the monument features ten embedded white marble reliefs that illustrate major

historical events in modern China.

人民英雄纪念碑由坚固美观的花岗石与汉白玉砌成（袁雪飞／摄）
The Monument to the People's Heroes made of granite and white marble (Photo by Yuan Xuefei)

梁思成与纪念碑的设计
Liang Sicheng and Design of the Monument

　　在天安门广场上建筑人民英雄纪念碑,具有重大社会意义。1949年9月30日纪念碑奠基典礼后,梁思成先生作为人民英雄纪念碑兴建委员会的副主任,主持了纪念碑的设计工作。设计团队中汇集了建筑师、艺术家、文史专家等,经过大量方案的比选和调整,最终形成了现在的纪念碑样式。1958年人民英雄纪念碑落成。

　　Construction of the Monument to the People's Heroes at the Tian'anmen Square is an event of great social significance. After the cornerstone laying ceremony on September 30, 1949, Professor Liang Sicheng, as vice chair of the Construction Committee, presided over the design of the Monument. The design work brought together architects, artists, historians and experts from other fields. After selection from quite a number of design proposals and due adjustment, the design plan was eventually decided. In 1958, the Monument to the People's Heroes was established.

人民英雄纪念碑须弥座处的汉白玉浮雕(顾彩华／摄)
White marble relief in the sumeru pedestal of the monument (Photo by Gu Caihua)

毛主席纪念堂
Chairman Mao Memorial Hall

毛主席纪念堂始建于1976年，平面呈正方形，占地面积5.72万平方米，总建筑面积近3.39万平方米，建筑高度33.6米。

纪念堂坐落于中轴线上，正门朝北，面对人民英雄纪念碑。

纪念堂建筑从中国传统木结构建筑中吸取灵感，以民族图案装饰立面。

Construction of the Chairman Mao Memorial Hall began in 1976. It has a square floor plan and covers an area of 57,200 square meters, with a floor space of 33,867 square meters and a height of 33.6 meters.

The structure lies on the Central Axis, facing the Monument to the People's Heroes in the north.

Its architectural design took inspiration from traditional Chinese wooden structures and its facade features patterns of national characteristics.

毛主席纪念堂北立面 (牛飞／摄)
Facade of Chairman Mao Memorial Hall (Photo by Niu Fei)

中国国家博物馆
National Museum of China

中国国家博物馆位于天安门广场东侧，坐东朝西，占地面积7万平方米，建筑面积近20万平方米。
建筑设计采用了廊柱庭院方式，三个内院构成"目"字形平面，和广场对面的人民大会堂形成一虚一实的对比。
博物馆建筑物的高度、长度和人民大会堂相近，
这样就从体量上保证了天安门广场两侧建筑以北京中轴线左右对称的构想。

The National Museum of China is situated on the east side of the Tian'anmen Square,
it faces west and covers an area of 70,000 square meters, with a floor space of about 200,000 square meters.
The structure is composed of three courtyards surrounded by porch columns,
forming a contrast with the Great Hall of the People on the opposite of the Tian'anmen Square.
Its height and length are similar to the Great Hall of the People,
ensuring symmetric designs and corresponding volumes of the two buildings on both sides of Beijing Central Axis
that runs through the middle of the Tian'anmen Square.

中国国家博物馆西立面（绿汀／摄）
The west facade of the National Museum of China (Photo by Lyu Ting)

中国国家博物馆庭院 (金东俊／摄)
Courtyard in the National Museum of China (Photo by Jin Dongjun)

中国国家博物馆入口空间宽敞而明亮 (金东俊／摄)
Spacious and well lighted hall of the National Museum of China (Photo by Jin Dongjun)

人民大会堂
Great Hall of the People

人民大会堂始建于1958年,位于天安门广场西侧,坐西朝东,建筑东面正门正对广场。

人民大会堂占地面积15万平方米,建筑面积17.18万平方米;

建筑造型平面对称,高低结合,台基、廊柱、屋檐均采用中国传统建筑图案加以装饰;核心建筑高度46.5米,两翼高约30米。

The Great Hall of the People was constructed in 1958.

It lies on the west side of the Tian'anmen Square, with its main east gate facing the square.

It covers an area of 150,000 square meters and has a floor space of 171,800 square meters.

The structure features a symmetric plan, integrating different well-proportioned component buildings.

Its stylobates, colonnades, and eaves adopt decorative patterns from traditional Chinese architectures.

Its core building is 46.5 meters high, and its wings are 30 meters high.

人民大会堂与人民英雄纪念碑(陆岗/摄)
The Great Hall of the People and the Monument to the People's Heroes (Photo by Lu Gang)

人民大会堂
Great Hall of the People

人民大会堂是全国人民代表大会召开国事会议的地方，是国家最高的权力机关所在地。它由位于中央的万人大礼堂、北部宴会厅和南部人大常务委员会办公楼三部分组成。其中，万人大礼堂西面设主席台，可容300～500人的主席团，东面是可容纳近万人的扇形会场，共三层，主席团的会议室和工作室设在主席台的后面。宴会厅可以同时安排5000人的座席。人大常务委员会办公楼在建筑南端，是比较独立的部分，设有常务委员会议厅、国宾接待厅、宴会厅等。代表大会的小组会议室共60余组，布置在大礼堂的周围。

The Great Hall of the People is the seat of the country's sovereign organ where the National People's Congress convenes. It is composed of the Great Auditorium in the center, the State Banquet Hall in the north and the office building of the Standing Committee of the People's Congress of China in the south. The Great Auditorium has a dais in the west section that can accommodate 300 to 500 delegates. The east section is a fan-shaped, three-story hall that has a capacity of nearly 10,000 people. The Bureau's meeting room and office room are behind the dais. The State Banquet Hall can host up to 5,000 diners. The official building of the Standing Committee of the National People's Congress is a relatively independent section located at the southern edge of the building complex. There are also 60 meeting rooms around the auditorium.

人民大会堂立柱（陆岗／摄）
The marble column of the Great Hall of the People (Photo by Lu Gang)

人民大会堂内景（陆岗／摄）
The interior view (Photo by Lu Gang)

人民大会堂内的万人大礼堂（陆岗／摄）
The interior of the Great Auditorium (Photo by Lu Gang)

Worshipping Heaven
and Farming Ceremony

正阳门—永定门段落
Zhengyangmen Gate – Yongdingmen Gate Section

正阳门

天坛
┊
先农坛

外城诸大街与御道遗址 ———————————————— 永定门 ——————————

祈年殿（范炳远／摄）
Hall of Prayer for Good Harvests (Photo by Fan Bingyuan)

夕阳下的祈年殿（袁雪飞／摄）
Hall of Prayer for Good Harvests in the Sunset (Photo by Yuan Xuefei)

正阳门—永定门段落构成了北京中轴线南段,这里既是皇帝祭祀皇天和先农等神祇的重要空间,具有强烈的礼仪性质,又是进入古代都城的前导空间,途经的前门大街和天桥是昔日繁华的市井娱乐场所与商贸场所,具有强烈的世俗色彩和商业功能。皇家文化和民俗文化在这里相互交织,构成了该段落复杂而多样的文化传统,丰富了北京中轴线的多元内涵。

位于该段落北端的正阳门城楼和箭楼,曾是北京内城九门中的正南门。其中,正阳门城楼位于天安门广场之南,从这里向北可俯瞰天安门广场;正阳门箭楼位于其城楼以南,从那里向南可俯瞰前门大街北段,并遥望永定门城楼。

天坛位于北京中轴线南段的东侧、北京外城的东南隅,是明清两代帝王祭祀上天和祈求丰收的场所。天坛从选址、规划、建筑的设计以及祭祀礼仪和祭祀乐舞,无不依据中国古代阴阳、五行等学说,成功地把古人对"天"的认识、"天人关系"以及对上苍的愿望表现得淋漓尽致。中国各朝各代均建坛祭天,而天坛是完整保存下来的仅有的一例,是中国古人的杰作。

先农坛位于北京中轴线南段的西侧、北京外城的西南隅,是明清两代皇帝祭祀先农诸神、太岁诸神和举行亲耕的地方。先农坛见证了中国古代农业社会所特有的祭祀农业神祇的文化传统,同样体现了中国古人的传统价值观和宇宙观,承载着明清时期国家礼仪祭祀传统。

前门大街、天桥南大街与永定门御道遗址既是明清时期重要的礼仪仪式行进路线,也是外城繁荣的商业街市。时至今日,前门大街、天桥南大街仍然是北京老城内重要的商业区、演艺区,数百年间延续了始建时的历史功能与文化传统。

永定门是明清北京外城正南门,在外城七门中,永定门的规制最高。永定门是北京中轴线的南端点,在城楼上可与北面的正阳门远远相望,是北京中轴线南段独特的景观节点。

The section between the Zhengyangmen Gate and the Yongdingmen Gate is the southernmost part of Beijing Central Axis. It marked an important space where the emperor offered sacrifices to heaven and various deities such as the God of Agriculture and at the same time, it was a space leading to the entrance of the ancient capital city. Qianmen Street and Tianqiao, where this section of the axis passed through, were bustling commercial and entertainment areas for commoners, bearing strong secular overtones.

The Zhengyangmen Gate and its Archery Tower, standing at the section's northernmost edge, was the main south gate among the nine gates to the Inner City. The Zhengyangmen Gate marks the southern boundary of the Tian'anmen Square, where one can overlook the entire square in the north. Its Archery Tower lies to the south of the gate, offering a look at Qianmen Street in the south and the Yongdingmen Gate in the distance.

The Temple of Heaven lies on the east side of the southern section of Beijing Central Axis and at the southeast corner of Beijing's Outer City. In the Ming and Qing dynasties, it was a place where emperors offered sacrifices to heaven and pray for good harvests. Its location selection, layout planning, architectural design as well as sacrificial ceremonies and music and dance performances, all were inspired by the traditional Chinese doctrines of Yin - Yang and the Five Elements, fully demonstrating ancient people's understanding of "heaven" and "relationship between man and nature" as well as their aspirations for blessings of heaven. Altars were constructed to offer sacrifices to heaven in all dynasties in ancient China, but the Temple of Heaven is the only example that survives to this day, representing an architectural masterpiece by ancient Chinese.

The Altar of Agriculture was where emperors of the Ming and Qing dynasties offered sacrifices to gods of agriculture and guardian gods of the year, and held ceremonies of land tilling. It has witnessed the cultural tradition of worshipping gods of agriculture that is unique to the farming society in ancient China. It also manifests the ancients' values and view of the universe and testifies to the state's tradition of sacrificial rituals.

Qianmen Street, Tianqiao South Street and the Yongdingmen Imperial Road Site formed an important route for imperial ceremonies during the Ming and Qing period. The area is also a bustling marketplace in the outer city. Today, Qianmen Street and Tianqiao South Street are still major commercial and performance areas within Beijing's old city, which have carried through hundreds of years their orignal functions and cultural traditions.

The Yongdingmen Gate was the main south gate to the Outer City in the Ming and Qing dynasties. It ranks the highest among all the seven gates of the Outer City, marking the southern terminus of Beijing Central Axis.

正阳门

Zhengyangmen Gate

正阳门城楼与箭楼（李彦成主编：《中轴旧影》，文物出版社）
Zhengyangmen Gate Tower and Archery Tower (*The Old Photos of the Central Axis of Beijing* edited by Li Yancheng, Cultural Relics Publishing House)

　　正阳门始建于明永乐十七年(1419年)，由城楼、瓮城与箭楼构成。它是明清两代北京内城的正南门、"京师九门"之一，其营建位置体现了《周礼·考工记》的规划理念。在北京内城九门、外城七门中，正阳门规模最为宏大，建筑规制等级最高，被视为都城之"国门"。

　　The Zhengyangmen Gate, first constructed in 1419 (the 17th year of the Yongle reign, Ming Dynasty), comprises the gate tower, the barbican walls and the archery tower. It was the main south gate of the inner city during the Ming and Qing dynasties. It was considered as the "National Gate", for its magnificent scale and the highest-ranking architectural specifications.

城

夜色中的正阳门城楼及箭楼（王心超／摄）
Tower and the archery tower of the Zhengyangmen Gate at night (Photo by Wang Xinchao)

正阳门箭楼

Archery Tower of the Zhengyangmen Gate

正阳门箭楼面阔七间，上下共四层，楼高24米。
城台高12米，门洞位于城台正中，
是北京内城九门中唯一箭楼开门洞的城门。楼身东、南、西三面开箭窗94个，供对外射箭用。

The archery tower of the Zhengyangmen Gate is a seven-bay, four-story structure, totaling 24 meters in height.
Its platform is 12 meters high, with the archway in the middle.
Among all the nine gates of the Inner City, the Zhengyangmen Gate is the only one that has an archway in its archery tower.
The archery tower has 94 arrow shooting windows on its east, south and west sides.

箭楼（绿汀／摄）
The archery tower of the Zhengyangmen Gate (Photo by Lyu Ting)

前门夜色（路士跃／摄）
Qianmen at night (Photo by Lu Shiyue)

正阳门城楼
Zhengyangmen Gate Tower

正阳门城楼为两层建筑，面阔七间，进深三间，周围设廊。

两层城楼均四面有门，上下两层四面均对开。

整个正阳门城楼占地面积3047平方米，建在高14.7米的城台上，城台南北上沿各有1.2米高的女墙。

The Zhengyangmen Gate Tower is a two-story structure that has seven bays in width and three bays in depth,
surrounded by corridors.
Each storey has gates on its four sides.
The whole structure stands on a 14.7-meter-high platform.

正阳门（袁雪飞／摄）
The Zhengyangmen Gate (Photo by Yuan Xuefei)

位于正阳门的『中国公路零公里点』标志牌
（陆岗／摄）
Marker of "Zero Point of Highways, China" at
Zhengyangmen Gate (Photo by Lu Gang)

正阳门城楼建在高14.7米的城台之上（陆岩／摄）
The Zhengyangmen Gate Tower was built on a
14.7-meter-high platform (Photo by Lu Yan)

从正阳门大街上北望正阳门五牌楼与箭楼（李响／摄）
The Zhengyangmen archway and archery tower viewed from the south (Photo by Xu Yan)

天坛
Temple of Heaven

天坛全景图，清晰可见天坛沿"皇乾殿—祈年殿—皇穹宇—圜丘"轴线对称分布的建筑格局 (李彦成主编:《中轴旧影》，文物出版社)
A panoramic view of the Temple of Heaven clearly showing the architectural layout symmetrically arranged along the axis of "the Hall of Imperial Zenith, the Hall of Prayer for Good Harvests, the Imperial Vault of Heaven, and the Circular Mound Altar" (*The Old Photos of the Central Axis of Beijing* edited by Li Yancheng, Cultural Relics Publishing House)

　　天坛始建于明永乐十八年(1420年)，整个建筑群由内坛和外坛两大区域组成。内坛主要建筑群包括承担"天地合祀"和冬至"祭天"功能的圜丘坛建筑群，承担孟春"祈谷"功能的祈谷坛建筑群，以及斋宫建筑群等；外坛设有神乐署建筑群。天坛选址体现了中国古代传统中"南郊祭天"的布局思想。

　　The Temple of Heaven was first constructed in 1420 (the 18th year of the Yongle reign, Ming Dynasty). The entire building complex consists of two areas: the inner altar and the outer altar. The inner altar mainly includes the Circular Mound Altar complex to host sacrificial rites to both heaven and earth collectively and to heaven individually on the Winter Solstice, the Altar of Prayer for Bumper Crops complex to host sacrificial rites in spring and the Palace of Abstinence where the emperor fasted before each ceremony. The outer altar contains the Divine Music Administration complex that offers music and dancing activities for sacrificial ceremonies. The location of the Temple of Heaven embodies the traditional planning idea of "worshipping heaven in the southern suburb".

天坛与明清皇帝
Temple of Heaven and Ming and Qing Emperors

　　天坛，明初称为天地坛，现有名称得于明嘉靖改制，嘉靖皇帝是最热衷于在天坛举行祭祀活动的明朝皇帝。最早在北京修建大祀殿的永乐皇帝进行过4次天地合祀仪式；明朝最后一个皇帝崇祯在位期间，进行过2次孟春祈谷，5次冬至祭天，2次祈雨雪、谢雨雪仪式。

　　相较于明代，清朝皇帝在天坛进行的祭祀活动整体上更加频繁。明朝皇帝在天坛共举行过149次祭祀活动，而清朝入关后的9代皇帝则举行了524次之多。其中乾隆皇帝是最热衷的一位，他在位期间进行了58次孟春祈谷，59次冬至祭天，38次常雩礼，祭祀次数空前绝后。

　　Ming emperors hosted a total of 149 sacrificial rites at the Temple of Heaven. After 1644 the Manchurian army conquered Beijing and made it the capital of the Qing empire, all the nine Qing emperors hosted 524 sacrificial rites at the Temple of Heaven. Emperor Qianlong, who was particularly keen on such ceremonies, hosted 155 sacrificial rites there.

圜丘坛棂星门(杜启法／摄)
Lattice Star Gates at the Circular Mound Altar (Photo by Du Qifa)

祈年殿(李少白／摄)
Hall of Prayer for Good Harvests (Photo by Li Shaobai)

圆丘坛 (董亚力/摄)
Circular Mound Altar (Photo by Dong Yali)

圜丘坛
Circular Mound Altar

圜丘坛为圆形汉白玉须弥座石坛,共分上、中、下三层,通高5.17米, 最上层坛中心设有天心石。
圜丘四面都有台阶,各层台阶都是九级。
圜丘四周环绕着汉白玉围栏,各层栏板望柱及台阶数目均用阳数,符合"九五"之尊的规制。

The Circular Mound Altar is a round stone altar of three levels on a white marble Sumeru Seat,
totaling 5.17 meters in height. The top layer has the Heavenly Center Stone at its center.
The altar has staircases on all sides leading to its uppermost terrace,
with nine steps through each level, and is encircled by white marble railings.

圜丘天心石（董亚力／摄）
Heavenly Center Stone at the Circular Mound Altar
(Photo by Dong Yali)

燔柴炉与燎炉（董亚力／摄）
Firewood Stoves and an Offering Burner (Photo by Dong Yali)

皇穹宇
Imperial Vault of Heaven

皇穹宇位于圜丘坛北面，建筑坐北朝南。
整个殿宇外观似圆亭，坐落在2米高的汉白玉须弥座台基上，周围设有石护栏。
皇穹宇为镏金宝顶单檐蓝瓦圆攒尖顶，直径15.6米，高19.2米，
由8根金柱和8根檐柱共同支撑起殿顶，有天花藻井三层，层层收进。

The Imperial Vault of Heaven is located north of the Circular Mound Altar, facing the south.
The whole structure is shaped like a round pavilion, sitting at a 2-meter-high marble Sumeru platform surrounded by stone guardrails.
It features a single-eave pyramidal roof decorated by a gilt top, with a diameter of 15.6m and a height of 19.2m.
The roof is supported by eight gilt pillars and eight eave pillars, featuring a caisson ceiling in three successive levels of diminishing size.

皇穹宇是供奉祭祀神位的场所 (范炳远／摄)
The Imperial Vault of Heaven is a place for enshrining tablets of worshipped gods (Photo by Fan Bingyuan)

皇穹宇 (站酷／供图)
Imperial Vault of Heaven (Provided by Zcool.com.cn)

天坛的第一次公众开放
First Public Access to the Temple of Heaven

1913年元旦，天坛曾经短暂地向公众开放。"中华民国二年一月一日为始，前门外天坛先农坛，各开放十日，任人入内游览。"这次开放满足了公众对于皇家祭天建筑游览的渴望，大量游客涌入天坛，"天坛门首，但见一片黑压压，人山人海，好像千佛头一般，人是直个点的往里灌"。游客中不仅有北京城的居民，甚至还有来自周边天津、保定的人。虽然开放仅持续10天，却标志着天坛从皇家祭坛转向公众活动空间迈出了第一步。1918年，天坛正式作为天坛公园对民众永久开放。

On New Year's Day of 1913, the Temple of Heaven was opened to the public for the first time in its history, which however lasted only for 10 days. It marked the first step forward for the Temple of Heaven being transformed from an imperial sacrificial altar to a place for public activities. In 1918, the Temple of Heaven was officially opened to the public as a park.

丹陛桥北通祈年殿(徐晉／摄)
The God Road leads to the Hall of Prayer for Good Harvests in the north (Photo by Xu Shen)

与北京中轴线平行的天坛中轴线
Central Axis of the Temple of Heaven: A Defining Axis in Parallel with Beijing Central Axis

天坛建筑群有自己的建筑轴线，在建筑中心偏东，为南北向，与北京中轴线平行。其建筑轴线上坐落着天坛两个主要的建筑群，南面是圜丘坛，北面是祈谷坛，两者由丹陛桥相连。

The Temple of Heaven complex has its own central axis. It is a south-north axis through the center of the complex by east, in parallel with Beijing Central Axis. On this central axis are two main building complexes of the Temple of Heaven: the Circular Mound Altar in the south and the Altar of Prayer for Bumper Crops in the north which are connected by the God Road.

丹陛桥
God Road

祈谷坛在北，圜丘坛在南，

两者之间由一条长360米、宽30米、向北缓缓升高的砖砌甬道即丹陛桥相连。

丹陛桥体现了天坛建筑群也有自身中轴线的特征。

The God Road is a 360-meter-long,

30-meter-wide brick passage slightly ascending northward,

connecting the Altar of Prayer for Bumper Crops in the north and the Circular Mound Altar in the south.

The God Road shows that the Temple of Heaven Complex has a central axis of itself.

祈年殿（董亚力／摄）
Hall of Prayer for Good Harvests (Photo by Dong Yali)

祈年殿
Hall of Prayer for Good Harvests

祈年殿坐落于圆形汉白玉台基即祈谷坛之上，

坛分三层，坛台通高5.2米，占地5900多平方米，每层都有雕花的汉白玉栏杆围绕。

祈年殿建筑总高38米，木结构直径24.2米，三层连体的殿檐由28根楠木大柱和36根互相衔接的枋桷支撑；

屋顶是镏金宝顶蓝瓦三重檐攒尖顶。殿内大柱分为内、中、外三层，

内层有4根龙井柱，中层为12根金柱，最外层是12根檐柱。

The Hall of Prayer for Good Harvests sits on a round white marble platform known as the Alar of Prayer for Bumper Crops.

The altar has three levels, covering an area of 5,900 square meters, with an overall height of 5.2 meters.

Each level is encircled by carved white marble railings.

祈年殿的圆顶象征天圆，蓝色琉璃瓦象征蓝天
（范炳远／摄）
The round roof of the Hall of Prayer for Good Harvests
symbolizes the round heaven while its blue glazed tiles
represent the blue sky (Photo by Fan Bingyuan)

祈年殿藻井（董亚力／摄）
Caisson Ceiling in the Hall of Prayer for Good Harvests
(Photo by Dong Yali)

皇乾殿坐落于祈年殿以北,通过三座琉璃门相通 (牛飞／摄)
The Hall of Imperial Zenith is located north of the Hall of Prayer for Good Harvests and with three glaze-tile gates standing in between (Photo by Niu Fei)

北宰牲亭 (董亚力／摄)
North Pavilion for Preparing Animal Sacrifices (Photo by Dong Yali)

先农坛
Altar of Agriculture

雍正帝祭先农坛图 (局部图; fotoe.com/供图)
Painting of Emperor Yongzheng Performing Sacrifices at the Altar of Agriculture (Partial graph; Provided by fotoe.com)

先农坛始建于明永乐十八年(1420年),明代一度称"山川坛"。先农坛整体布局与天坛类似,呈现北圆南方的格局,占地面积约130万平方米,由内外两层围墙环绕。先农坛的内坛是祭祀太岁和先农的核心场所,由三组重要的祭祀设施构成,包括太岁殿建筑群、先农坛、观耕台和耤田,以及为了祭祀而准备的设施,比如神仓建筑群、神厨建筑群、具服殿等。

The Altar of Agriculture was first constructed in 1420 (the 18th year of the Yongle reign, Ming Dynasty). Its inner altar is the core area to offer sacrifices to gods of agriculture and gods of the year. It is composed of three major groups of sacrificial facilities, including the Hall of the God of the Year complex, the Altar of Agriculture, and the Platform for Viewing Plowing and Sacred Fields, as well as facilities for preparing sacrificial rites.

观耕台及皇帝亲耕的一亩三分地（金东俊／摄）
The Platform for Viewing Plowing and the Imperial Cultivated Field (Photo by Jin Dongjun)

　　先农坛的选址、规划以及祭祀礼仪等方面见证了中国古代农业社会所特有的祭祀先农的文化传统，体现了中国古人的传统价值观和独特宇宙观。先农坛将中国古代的礼制思想与建筑规划设计思想以及方位观、时空观等思想学说结合在了一起。先农坛与天坛分列北京中轴线两侧，强化了北京中轴线的规划格局。

　　The location selection and planning of the Altar of Agriculture as well as related sacrificial rituals are all witnesses to the cultural tradition of worshipping gods of agriculture that is unique to the farming society in ancient China and manifests the ancient Chinese values and outlook of the universe. The Altar of Agriculture and the Temple of Heaven, symmetrically arranged on both sides of Beijing Central Axis, are the reinforced presentations of the symmetric layout of the axis.

观耕台
Platform for Viewing Plowing

观耕台是皇帝亲耕完毕后观看王公大臣们耕作的高台。

观耕台为方形，边长18米，高1.9米，东、南、西各有台阶9级，台阶踏步由汉白玉条石砌成，侧面雕有莲花图案。

台上四周有汉白玉石雕栏板，台底须弥座由黄绿琉璃砖砌筑，琉璃砖上雕刻了龙缠草图案。

The Platform for Viewing Plowing was a high platform where the emperor viewed farming activities of nobles and officials after he completed his own ceremonial farming practices.

观耕台（站酷／供图）
The Platform for Viewing Plowing (Provided by Zcool.com.cn)

先农坛
Altar of Agriculture

皇帝祭祀先农神的祭坛
位于内坛西北，坐北朝南，是一座砖石结构的方形平台，
边长15米，坛高1.5米，四面有台阶各8级。

The Altar of Agriculture lies in the northwest of the inner altar area, facing south.
It is a square platform of brick-and-stone structure, with a side length of 15 meters and a height of 1.5 meters.
It has staircases on four sides, each with eight steps.

先农坛台（欧阳平／摄）
The Altar of Agriculture (Photo by Ouyang Ping)

太岁殿
Hall of the God of the Year

太岁殿位于先农神坛东北，内坛北门西南侧，
是祭祀太岁神及十二月将神的院落。
它居内坛建筑的中心地带，建筑体量是先农坛中最大的，其木构架结构形式基本与故宫太和殿上层类似。
The Hall of the God of the Year is located in the northeast of the Altar of Agriculture and
the southwest of the north gate to the inner altar.
It is a courtyard structure where the God of the Year and deities guarding twelve months of the year are worshipped.
Standing at the center of the inner altar area,
the Hall of the God of the Year is the largest among all the buildings of the Altar of Agriculture.

太岁殿建筑体量为先农坛之最 (杨春燕／摄)
The Hall of the God of the Year is the largest among all the buildings of the Altar of Agriculture (Photo by Yang Chunyan)

太岁殿外檐平身科斗拱 (欧阳平／摄)
The Dougong on the outer eave of the Hall of the God of the Year
(Photo by Ouyang Ping)

拜殿金龙和玺彩绘与斗拱 (欧阳平／摄)
The Golden Dragon painting and the Dougong of the Worship Hall
(Photo by Ouyang Ping)

地祇坛遗存(杨春燕 摄)
The remains of the Altar of Earthly Gods (Photo by Yang Chunyan)

神祇坛
Altars of Heavenly and Earthly Gods Site

神祇坛是天神坛、地祇坛的合称,始建于明嘉靖十一年(1532年),原位于先农坛内坛南门外,两坛四周各有青石棂星门及矮墙环绕。东为天神坛,坐北朝南,坛台北面有四座云海纹石龛,供奉云、雨、风、雷之神。

西为地祇坛,坐南朝北,坛台东、西、南三面共设九座山岳、江海纹石龛,供奉天下名山大川之神。

神祇坛现仅存遗址,坛台已无存。

The Altars of Heavenly and Earthly Gods, first constructed in the 11th year of the Jiajing reign, Ming Dynasty (1532), is a collective term for the Altar of Heavenly Gods and the Altar of Earthly Gods. The Altar of Heavenly Gods in the east and facing south has four carved stone niches where the gods of cloud, rain, wind and thunder were enshrined.

The Altar of Earthly Gods in the west and facing north has nine stone niches where gods of mountains and rivers were enshrined.

The original buildings of the Altars have been lost.

地祇坛四海石龛（欧阳平／摄）
Stone niches of four seas (Photo by Ouyang Ping)

地祇坛山岳石龛（欧阳平／摄）
Stone niches of mountains (Photo by Ouyang Ping)

庆成宫
Qingcheng Hall

庆成宫建筑群位于外坛东侧，初为斋宫，是祭典前皇帝进行斋戒的地方。乾隆二十年（1755年）改今名，
皇帝亲耕后会在此行庆贺礼并赐茶。
正殿位于庆成宫院落中轴线北部，建筑建于台基之上，殿前有月台，月台南侧还有日晷石座和时辰碑石亭。
庆成宫建筑群和太岁殿建筑群形成两组南北向轴线对称的布局，轴线与北京中轴线平行，
它们与耤田一同展现了独特的景观。

Qingcheng Hall complex is located in the east of the outer altar.
It was the place where the emperor fasted before the sacrificial ceremony in the Ming and early Qing dynasties.
After 1755 (the 20th year of the Qianlong reign), the emperor celebrated the success of the ceremony in this hall.

庆成宫（牛飞／摄）
Qingcheng Hall (Photo by Niu Fei)

外城诸大街与御道遗址

Streets in the Outer City and the Imperial Road Site

昔日前门大街西侧店铺（李彦成主编：《中轴旧影》，文物出版社）
Shops on the west side of Qianmen Street (*The Old Photos of the Central Axis of Beijing* edited by Li Yancheng, Cultural Relics Publishing House)

　　前门大街、天桥南大街与永定门御道由北向南延伸，连接了北京内城与外城的交通，这些道路曾经是明清帝王进行祭礼活动、出入京城的礼仪线路，同时也是外城繁荣的商业街市。

　　Qianmen Street, Tianqiao South Street and Yongdingmen Imperial Road run through Beijing Central Axis and extend southward, connecting the inner city and the outer city. These roads offered a route for Ming and Qing emperors to march from the Palace City to sacrificial altars. They were also bustling marketplaces in the outer city.

前门大街 (站酷／供图)
Qianmen Street (Provided by Zcool.com.cn)

前门五牌楼和正阳门箭楼 (李玲／摄)
The Archway and the archery tower of the Zhengyangmen Gate (Photo by Li Ling)

前门大街
Qianmen Street

前门大街北自正阳门起始，南至天桥，在明清时期至20世纪初皆称正阳门大街，因正阳门是北京内城正门，也被称为前门，所以民众将该大街俗称为前门大街。今天，前门大街仍然是北京重要的商业区与文化休闲场所。

Qianmen Street starts from the Zhengyangmen Gate in the north and ends at Tianqiao in the south. It was called "Zhengyangmen Street" from the Ming and Qing dynasties until the early 20th century and was later commonly known as "Qianmen Street".

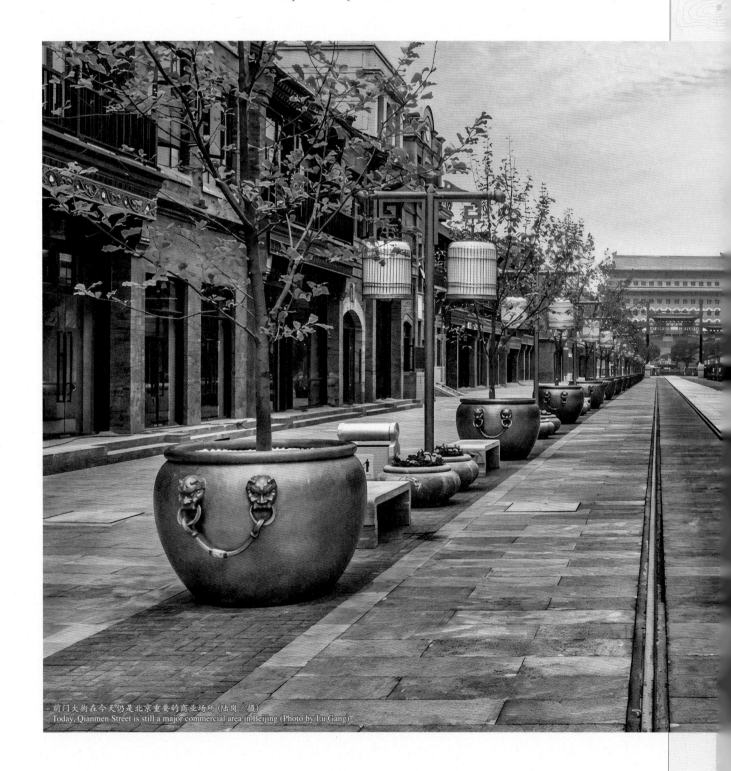

前门大街在今天仍是北京重要的商业场所 (陆岗／摄)
Today, Qianmen Street is still a major commercial area in Beijing (Photo by Lu Gang)

天桥南大街
Tianqiao South Street

天桥南大街在前门大街以南，
北自天桥，
南至现今永定门广场北端。
To the south of Qianmen Street is Tianqiao South Street that
stretches from Tianqiao in the north to
the north end of the present-day Yongdingmen Square in the south.

位于天桥处的北京中轴线地面标识(范炳远／摄)
The floor mark of Beijing Central Axis at Tianqiao (Photo by Fan Bingyuan)

天桥——百年来平民娱乐之所
Tianqiao, an Entertainment Place for Commoners

　　天桥地区最为繁荣的阶段为民国初期，由于前门外荷包巷的集市被拆除，商贩们便转移到天桥两侧的空地上开设店铺。后来，进京闯荡的江湖艺人也聚集于此卖艺，各路戏园陆续搬迁而来扎堆演出，天桥由此成为全城著名的民间娱乐场所。新中国成立后，天桥地区又新建剧场、影院、自然博物馆等，2013年建设有天桥艺术中心，进一步强化了这里的娱乐、民俗和教育功能。

　　The Tianqiao area saw its heydays during the early 20th century when buskers gathered and theatres moved here. Tianqiao thus became the most popular place for folk entertainment in Beijing. Since the 1950s, new theatres, cinemas and museums have been set up here.

永定门御道遗址 (范炳远 / 摄)
The Yongdingmen Imperial Road Site (Photo by Fan Bingyuan)

永定门御道遗址
Yongdingmen Imperial Road Site

永定门御道遗址是在修建永定门广场时通过考古发掘进行揭露展示的一段道路遗址，
南北长约200米，宽约13米，
为花岗岩石条铺设。

The Yongdingmen Imperial Road Site is a section of road site revealed
by archaeological excavation when the Yongdingmen Square was constructed.
It is an archaeological site paved by granites slabs, totaling 200 meters in length and 13 meters in width.

由永定门望北京中轴线南段(陆建成／摄)
View over the south section of Beijing Central Axis from Yongdingmen Gate (Photo by Lu Jiancheng)

永定门
Yongdingmen Gate

永定门瓮城（李彦成主编：《中轴旧影》，文物出版社）
The Barbican of the Yongdingmen Gate (*The Old Photos of the Central Axis of Beijing* edited by Li Yancheng, Cultural Relics Publishing House)

　　永定门始建于明嘉靖三十二年（1553年），作为昔日北京外城的正南门，是明清两代城市防御体系和城市管理设施的重要组成，因现代交通建设于20世纪50年代被拆除。现永定门地标作为北京中轴线南端点的地理标识于2005年复建完成，由城楼建筑及瓮城地面标识构成，现已成为北京中轴线南段民众休闲活动及登高望远的公共空间。

　　The Yongdingmen Gate was first constructed in 1553 (the 32th year of the Jiajing reign, Ming Dynasty). As the main south gate to the outer city, it is an important part of the city defense system and city management facility network during the Ming and Qing dynasties. It was removed in the 1950s to make way for the development of transportation. The Yongdingmen Gate Tower was reconstructed in 2005.

北京中轴线南端点 (阮旭红／摄)
Southern terminus of Beijing Central Axis (Photo by Ruan Xuhong)

永定门门钉 (范炳远／摄)
Decorative nails of the Yongdingmen Gate (Photo by Fan Bingyuan)

复建的永定门城楼及广场是市民登高望远与休闲活动的场所(范炳远／摄)
The reconstructed Yongdingmen Gate Tower and Square are now leisure space for citizens (Photo by Fan Bingyuan)

永定门外燕墩(范炳远／摄)
Yandun Guardian Platform outside the Yongdingmen Gate (Photo by Fan Bingyuan)

向北穿过永定门，就进入了老北京城(金东俊／摄)
Passing through the Yongdingmen Gate, one would enter Beijing's old city in the north (Photo by Jin Dongjun)

永定门御道遗址 (范炳远／摄)
Yongdingmen Imperial Road Site (Photo by Fan Bingyuan)

后记
Postscript

在谈论古代建筑的时候，人们会说建筑是石头的史书。这些建筑反映了建造它们的时代和建造它们的人的生活和思想，它们的身上承载着岁月的印记，成为人们认识和理解历史发展的物质媒介。

北京是一座有着悠久历史的城市，它的建成年代可以上溯到公元前1000年。在13世纪忽必烈将元朝的首都选定在今天北京老城的位置，从此翻开了北京作为中国统一国家首都的篇章。在漫长的近八个世纪中，这里发生着影响了中华文明发展方向的事件，产生了许多使中华文明焕发出光辉的思想，更有无数的人在这里走过，留下他们的足迹，留下生活的悲喜故事。

北京中轴线是这一连串故事的空间和时间的经纬脉络。穿过昔日宏伟的红墙黄瓦的宫殿、古树森然的帝王苑囿和坛庙，听钟鼓楼悠远的暮鼓晨钟、阵阵鸽哨，观正阳门绕楼飞翔的雨燕，漫步跨越古老运河雕饰精美的古桥、繁华的市井，驻足壮丽的城市广场，北京中轴线展现了一幅蕴含着传统礼乐精神的古今交融的图景。这一切无声的讲述，都在促使人们去探索和理解今天这座城市当代繁华背后悠远的故事。

北京中轴线上的建筑群和城市空间决定了这座城市在古代的布局形态和未来的发展。它是北京的缩影，也是中华文明理想世界的呈现。

感受北京中轴线，感受北京城，感受这座城市所表达的文明精神。

When it comes to ancient architecture, it is commonly acknowledged that architecture is a history of the stone. Ancient architecture not only reflects the times when they were built as well as the lives and thoughts of the people who built them, but also bears the imprint of time that evolved into the material medium for people to cognize and comprehend the course of history.

Beijing is a city with a long history, which dates back to the year 1000 BC. In the 13th century, Kublai Khan gave orders to build the capital of the Yuan Dynasty on the location of the old city of today's Beijing, which made a new start for Beijing as the capital of a unified country in Chinese history. During the long period of recent eight centuries, remarkable events have taken place here that had a far-reaching influence on the heading of Chinese civilization, with the emergence of great thoughts that rejuvenate this civilization and numerous people leaving their footprints and stories of sweetness and bitterness in life.

Beijing Central Axis is the spatial and temporal context of this series of stories. Beijing Central Axis manifests a scene of integrating the ancient and modern times that contains the spirit of traditional rites and music, which can be reflected in the magnificent palaces with red walls and yellow tiles, the imperial gardens and temples with dense ancient trees, the remote morning bell and evening drum of the Bell and Drum Towers, the whistles of pigeons, the swifts hovering around Zhengyangmen Gate, the exquisitely carved ancient bridges over the ancient canals, the bustling marketplace, and the magnificent city square. The silent narration is spurring people to explore and perceive the long-standing story behind the contemporary prosperity of this modern city.

The architectural complex and urban space on Beijing Central Axis determine the ancient layout and future development of the city which is not only the epitome of Beijing but also the manifestation of the ideal world of Chinese civilization.

The journey will be taken to feel Beijing Central Axis, feel the city, and feel the essence of the civilization.

图书在版编目（CIP）数据

天地中和 ：北京中轴线文化遗产 ：汉、英／吕舟
主编． — 北京 ：北京出版社，2023.1
ISBN 978-7-200-17309-3

Ⅰ．①天… Ⅱ．①吕… Ⅲ．①建筑物—介绍—北京—
汉、英 Ⅳ．① TU-862

中国版本图书馆 CIP 数据核字 (2022) 第 119350 号

项目策划：刘　扬
责任编辑：高　琪
责任印制：燕雨萌

天地中和
北京中轴线文化遗产
TIANDI ZHONGHE
吕舟　主编

*

北 京 出 版 集 团
　　　　　　　　　　出版
北 京 出 版 社

（北京北三环中路 6 号）

邮政编码：100120

网址：**www.bph.com.cn**

北京伦洋图书出版有限公司发行

北京雅昌艺术印刷有限公司印刷

*

889 毫米 ×1194 毫米　16 开本　13.75 印张　329 千字
2023 年 1 月第 1 版　2023 年 1 月第 1 次印刷
ISBN 978-7-200-17309-3

定价：998.00 元

如有印装质量问题，由本社负责调换

质量监督电话：010-58572393